Our World Ocean

Understanding the Most Important Ecosystem on Earth

Voyage I

Blue Planet Publishing
Fullerton, California

Our World Ocean

Written and Created by

Sean Chamberlin

With Illustrations by

Nicole Shaw

Cover and Page Art by

Martha Rich

Edited by

Denise Logsdon

Proofreading by

Christina Palaia

Page Design by

Amy Weiher

Blue Planet Publishing
Fullerton, California

The Voyages

Voyage I Ocean Science

Voyage II Ocean Geology & Chemistry

Voyage III Ocean Physics

Voyage IV Ocean Biology

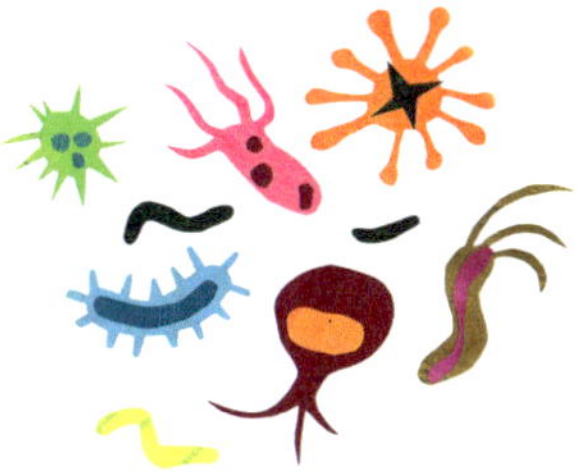

Voyage V Ocean Ecology

This book was inspired by Jacques Cousteau.

ISBN-13 978-1-7373677-2-7 (Pbk.)

Blue Planet Publishing
Fullerton, California

Dedicated to the people, animals, and robots helping to preserve our blue planet.

Blue Planet Publishing
Fullerton, California

Contents

Voyage I

Chapter 3 Science Aboard Ships

Chapter 4 Robots, Satellites, and Observatories

Chapter 5 Acoustical Oceanography

Appendices

What Students Say

Jill Craaybeek, Fall 2021

"Thank you for teaching me and opening my eyes to so many things."

Ryan Wren, Fall 2021

"I've always been interested in the ocean, but it's just so much more fascinating now with everything I've learned in this course."

Gabriel Hernandez, Fall 2021

"I learned a lot about the ocean that I was not aware of before! It has definitely inspired me to learn more about certain topics and to be more conscious about the environment in general."

Christina Classen, Spring 2022

"I'm so glad I remembered your name from a recommendation by a former student who took your oceanography class 20 years ago. . . . You made a big impression on him and that stuck with me all these years."

Vedania Chamorro, Spring 2022

"It was a tremendous wake-up call to me, in regards to how much I needed to change in my life so that I can help contribute to the health of our planet and every living creature on it."

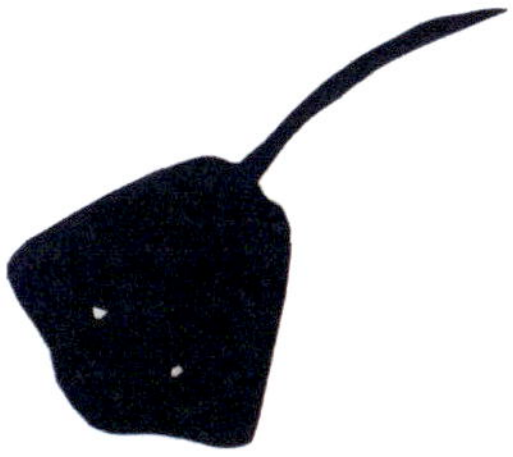

Serafin Héctor Junior, Spring 2022

"This class was very informative and I am glad I took it. Congratulations on your textbook."

Jena Pourschafie, Fall 2022

"Your care and encouragement are greatly appreciated. I did not feel stressed during your class, which was a very relaxing and calming virtual environment."

Credit: Chamberlin

Foreword

by Dr. Bridget Salzameda, Dean, Natural Sciences, Fullerton College

From the ground, a family sees my body clinging to the side of a mountain, about two-thirds of the way to the top. My arms are feeling weak as I grip two rocks wedged into the caked dirt. Thankfully my legs are firmly planted on two more wedged rocks. But I'm worried about my newly painted, bright red nails. OMG! What's wrong with me? Focus, Bridget. How did I get here, and what's my plan now?

I was forty-three years old and free solo climbing a mountain—really a shallow hill—with my climbing partner just a few feet across from me. I looked over, and he was gone. I looked down. He wasn't there. I called for him and heard no response. I assumed he was around the curve. He was, and climbed back down. I thought about returning to the bottom but badly wanted to reach the top because I was so close. I was tired, my arms were weakening, and I convinced myself to finish because I had witnessed a ten-year-old climb to the top an hour earlier. This made the task possible. If I hadn't seen it, I might not have continued. Civil and children's rights activist Marian Edelman's words bombarded my mind: "You can't be what you can't see." I made it to the top. I was out of breath, but content. Now I understood those words.

The first time I fell in love I was about eleven years old. I fell in love with my at-home chemistry kit. I was enamored by the brightly colored solutions evolving into different-looking substances in a thin, delicate test tube. My parents, who emigrated from Sri Lanka, gave me this toy, not realizing that it would help guide the trajectory of my career from student to chemistry professor to dean of natural sciences.

My wonder grew in my high school chemistry class in Mission Viejo, California. I was mesmerized by chemical reactions and reliable math problems and solutions. As a chemistry major at the University of San Diego, my uncomfortably direct academic advisor, Dr. Pat Traylor, called

me into her office and told me I must get a PhD in chemistry. I had no idea what a PhD was, but I was certain I would get one, because Dr. Traylor said I must. It was an important lesson in believing in others so they may believe in themselves. I worked hard in college and was accepted into several PhD programs, but I chose to earn my degree in biochemistry, not chemistry. Perhaps it was the rebel in me.

In graduate school, I met my muse. Dr. Christine Cremo, Department of Pharmacology, was my research advisor and a professor at the University of Nevada School of Medicine. She possessed the most important qualities that I wanted—the ability to produce and expect high-quality work, set high standards for herself and others, and demand rigor and intelligence in her research laboratory. She was organized, analytical, and direct. I was productive in the laboratory and blossomed as a research scientist, authoring scientific publications and making new discoveries to advance science. I earned my PhD and obtained the title of mother along the way. Others believed I would continue my research career and lead my own lab, but I had an internal calling to follow another path—to teach college chemistry (which I did for close to fifteen years).

Most would assume that the biggest challenge of being a mother and a professor is balancing work and family life. This is far from the truth. I have marvelous children, a supportive spouse, intelligent colleagues, and strong friendships. All of these, combined with efficient planning, allowed me to feel present in all I did at work and at home. This combination of skills, passion, and support is reflected in my most recent recognitions, receiving an Inspirational Professor Award and being a finalist for the Faculty of the Year award at our college.

But new challenges faced me in the last five years. Because my children were older, I took on more responsibilities at work, and as an administrator, I had time to pay attention. I was less naïve and less immune to blunt racism and microaggressions. I noticed my colleagues and friends experiencing it daily and began recognizing it from my past interactions. I am wholeheartedly thankful that I was naïve for my own emotional and mental protection—I had the luxury of focusing on science, not realizing there was anything standing in my way. As I work with students who are more aware of these topics, I am learning about the difficulties that underrepresented minorities experience. It's challenging to navigate my leadership role to ensure that all colleagues and students feel seen and heard. I understand the importance of surrounding myself with people who think differently so I can maintain a balanced perspective.

Representation is also important. All people deserve to see themselves in all areas of success, and science is no different. The outcome of my climb might have been different if I hadn't seen that ten-year-old completing it. This textbook will reach you differently because it incorporates lessons from a group of diverse scientists. The author followed his inner voice—probably the rebel in him—to thoughtfully communicate with students. It's not a boring textbook. It's fun, magical, and will allow you to learn and grow. You may find yourself smiling and smirking while reading this book, all while growing intellectually. Have fun falling in love!

Bridget Salzameda

Dr. Bridget Salzameda

Dean, Natural Sciences, Fullerton College

August 22, 2022

Credit: Chamberlin and Shaw

Preface

Aloha! If you're reading this, you're one of those rare people who read front matter. Perhaps you want to know what you're in for on this voyage of the sea. (An exhilarating one!) Maybe you want to know if we're your kind of people. (We hope so!) Possibly you're looking for the answer to a test question. (Keep reading.) Whatever your reason, we thank you. We hope you'll continue reading and enjoy your reading. After all, we can't protect what we don't understand.

When we set out to write this book (more than 12 years ago), Chamberlin had just finished publishing his first book, *Exploring the World Ocean*, with co-author and oceanographer extraordinaire Tommy Dickey (Chamberlin and Dickey 2008*). That book, published by McGraw-Hill, took the form of a traditional textbook. But there was a nagging feeling that, like so many textbooks, that textbook presented its subjects like bags of old bones. The meats and flavors were missing. Oceanography, like all science, is alive. It's alive! So Chamberlin set out to write a non-traditional, student-centered, science-infused book that inspired people to fall in love with the ocean and the people who have dedicated their lives to understanding how it works.

This book offers an alternative to the traditional, dry-as-bones, ho-hum approach of the standard textbook. Sure, it covers all the topics you'll find in a traditional oceanography textbook. But that's just the tip of the iceberg. We go broader and deeper. We write the way people talk—conversationally. We encourage folks to interact with the text, to click through to full-size images, to view video and hear audio, and to quickly access the reference materials on which the presented facts and ideas are based. We also offer an audio glossary! Just click on the blue bolded terms (PDF and ebook version only). You'll also find cartoons that tickle the

funny bones and imagery that makes you feel and think. The design makes it easy to read, gives ample space for posting notes, and numbered headings and subheadings make topics easier to find and make it easier for teachers to assign parts of chapters.

To our way of thinking, books do their best work when they stimulate action from the reader. As teachers well know, the strongest foundation for learning comes from attention to three domains: the affective (i.e., emotional), cognitive (i.e., intellectual), *and* psychomotor (i.e., physical). The structure, design, writing, artwork, and interactive elements of this book strive to engage those domains.

We also believe that science achieves its greatest success when it's built on the interdisciplinary and multiperspective works of diverse scientists. The problems of the world are best solved by giving everyone a voice in their solution. Scientific progress relies on the diverse knowledge, experience, and perspectives of scientists from all walks of life. That's why you'll find references with scientists' names throughout the text. We want readers to identify with names they recognize and develop curiosity about ones they don't. This approach builds a stronger appreciation for the contributions from scientists all over the world. We also make visible the dates of publication. Modern science is built on the shoulders of scientists spanning centuries of discovery. Making obvious the dates of publications connects readers to the history of science.

Of course, none of this matters if we bore our readers. We unapologetically use humor and irreverent analogies wherever possible to make our points. We've made every effort to make the text easy to read, the images engaging to look at, and the evidence for our statements simple to find. That doesn't mean we don't present difficult concepts and uncertainties in scientific knowledge. Readers deserve the "full monty" of science. How can the public develop an appreciation for the complex problems facing humanity if we treat them like dummies?

If you like what you read, please tell someone and share a copy. Our website version is 100% free (as long as you give appropriate credit and don't violate copyrights or CreativeCommons licenses). If you don't like what you read here, please tell us. We aim to do better.

May the ocean be with you. May a giant yellow sponge wipe away your tears when you're sad. May a thousand dolphins sing at your next big event. In the spirit with which this was written, aloha!

Sean, Nicki, and Martha

PS: The answer to the test question is C.

*__Reference:__ Chamberlin, W. Sean, and Tommy Dickey. 2008. *Exploring the World Ocean*. Dubuque: McGraw-Hill Higher Education.

Credit: istock.com/Michelle de Villiers

About This Book

Our World Ocean: Understanding the Most Important Ecosystem on Earth provides a survey of ocean science as it's practiced by oceanographers in the twenty-first century. The series aims to help readers understand the world ocean as a system of interdependent parts and processes. The series presents ocean science as a pursuit to know how the ocean works as a system.

The series is divided into five books. *Our World Ocean, Voyage I: Ocean Science* (5 chapters) covers the scientific method and the technologies oceanographers use to carry out 21st-century oceanographic research. *Voyage II: Ocean Geology and Chemistry* (6 chapters) presents the geology and chemistry of the ocean with an eye toward understanding climate change in the geologic past. *Voyage III: Ocean Physics* (9 chapters) explores the physics of the ocean and especially its role in the transport of heat in the Earth system. *Voyage IV: Ocean Biology* (17 chapters) dives into the biology of the ocean, from microbes to whales, highlighting the functional roles of organisms and their transfer of materials and energy within ecosystems. *Voyage V: Ocean Ecology* (8 chapters) brings it all together with a peek at ecological thinking about the ocean and human interactions with ocean ecosystems.

This five-book approach allows us to better serve a diverse community of readers and teachers. Bundled chapters permit us to modify and update smaller sets of materials and make them available more quickly than a single book with 45 chapters. Some folks won't be interested in all of the topics. So we figured a bundled approach would provide greater flexibility. That said, we're happy to bundle individual chapters together for your classroom. Call it a customized voyage. Otherwise, we're happy to help you use the book in every way possible. Feel free to reach out to sean@blueplanetpublishing.com.

How to Use This Book

This book provides a number of important tools for students, teachers, and ocean scientists interested in these topics. We designed it to be accessible, engaging, interactive, comprehensive, and understandable for all readers, no matter their previous experience with ocean science. If you discover something that would make the book better for you, by all means, contact us. We wrote it for you!

For Students

- one-column text for ease of reading on smartphones, tablets, laptops, and desktops (though we recommend viewing figures on as large a screen as possible)
- minimal technical jargon (wherever possible)
- clear explanation of concepts using everyday analogies
- use of humor and a strong authorial voice to grab your attention
- numbered and colored subheadings to make it easier to find and refer to specific sections of the text
- call-out boxes to highlight main points within the text
- hyperlinked in-text references to help you appreciate the millennial-long history and global diversity of ocean science
- wide margins for notetaking using Adobe Reader's comment tool or making notes on printed copies
- clickable colored and bolded terms linked to an audio glossary so you can hear definitions without having to leave the page
- abundant supporting photos, figures, and illustrations with alt text and side-by-side captions to engage you mentally and emotionally

For Teachers

- web version of the book for use in the classroom or on devices not ideal for PDFs
- figure credits with hyperlinks to all photos, illustrations, and graphics; public domain and Creative Commons photos (including ours) may be downloaded individually for use in teaching and presentations
- hyperlink access to in-text references for deeper exploration of a topic (and so you can check our interpretations, too!)
- availability (for a small price) of teaching and assessment materials, including PowerPoint and Keynote slides, test banks (Microsoft Word, Canvas, Blackboard, Respondus), alt text and figure captions (Microsoft Word), and more at TeachersPayTeachers. Your purchase of these materials makes it possible for us to create and distribute this book to the people who need it the most, students.

For Ocean Scientists

- a comprehensive survey of ocean science for getting up to speed on unfamiliar topics
- hyperlinks to references for quick access to the most recent publications on a particular research topic (granted, it's difficult to keep up, but we do our best, and if you can't find something, contact Chamberlin)
- free use of more than 500 cartoons on all fields of ocean science, ideal for that first slide at your next talk

Blue Planet Publishing
Fullerton, California

Coral reef in Sharm el Sheikh, Egypt. Credit: Ocean Image Bank/Renata Romeo

VOYAGE I

Ocean Science

A space oddity.

Artwork by Martha Rich.

The World Ocean System

1

Figure 1.1
The world ocean thrives as a system of interdependent parts and processes.

Credit: istock/ LeventKonuk

Imagine you are a cosmic explorer on a long, interstellar voyage between your solar system and its closest neighbor. Your trusty and comfortable spaceship has carried you billions and billions of miles from home, yet for most of that trip, you've seen very little. Sure, that asteroid almost hit you, and your onboard computer developed a split personality half a parsec back, but other than that, your trip through interstellar space has been one of the most boring experiences of your thousand-year life.

Then, out of the corner of your telescopic viewscreen, a solar system with a strange blue dot appears. You do a systems check. All systems normal. Suddenly the dot increases to two pixels. And then more. Switching to ultrasupercalifragilistic high definition, you realize that the dot is a real object. It's real! And it's BLUE. Could there be something, a water-rich object, in this ghastly stretch of nameless space? Phone home, explorer. You just found a water planet!

Viewed from the far reaches of space, Earth, our home, looks like a tiny blue speck immersed in a sea of black. We know Earth looks this way because in 1990 a small spacecraft, *Voyager 1*, going where no manmade

Figure 1.2
How many other water planets exist in space?

Credit: *Shaw and Chamberlin*

object had ever gone before, turned its camera back toward Earth and snapped a picture. In doing so, *Voyager 1* captured a moment that has come to define what it means to be human. Deeply moved by this image, Carl Sagan (1934–1996), astronomer, author, and host of the original 1980 TV miniseries *Cosmos*, offered these words:

> That's here. That's home. That's us. On it, everyone you ever heard of, every human being who ever lived, lived out their lives. The aggregate of all our joys and sufferings, thousands of confident religions, ideologies, and economic doctrines, every hunter and forager, every hero and coward, every creator and destroyer of civilizations, every king and peasant, every young couple in love, every hopeful child, every mother and father, every inventor and explorer, every teacher of morals, every corrupt politician, every superstar, every supreme leader, every saint and sinner in the history of our species, lived there—on a mote of dust, suspended in a sunbeam. (Sagan and Druyan 1994)

The **pale blue dot**, as Sagan called it, is Earth, our home, the only known place in the universe where humans exist. Though small compared to the infinity of space, Earth remains our lone island, our singular port of call, a solitary outpost inhabited by the most fascinating and enigmatic species in the known universe, *Homo sapiens*. Earth is also the only known planet with liquid water covering most of its surface. Its pale blue color, visible from millions of miles away, comes from that water, the ocean—our world ocean.

Figure 1.3
Viewed at a distance of 3.7 billion miles (6 billion km) from the Sun, Earth appears as a tiny blue speck bathed in rays of sunlight.

***Credit:* NASA/JPL-Caltech**

1.1 A World Ocean Perspective

The **world ocean** covers 71 percent of the surface of our planet to an average depth of 2.3 miles (3.7 km). It includes all of the waters contained by the ocean basins—which cover 64.7 percent of Earth's surface—and the waters that cover the edges of the continents—about 6.3 percent of the globe (e.g., Harris et al. 2014) . Myriad terms describe various parts of the ocean (seas and gulfs, straits and passages, bights and bays, harbors and sounds, banks and fjords), but **oceanographers**, the people, animals, and robots who study the world ocean, recognize only one ocean, the world ocean.

For convenience oceanographers call the three major ocean basins by their geographic names—the Atlantic, Pacific, and Indian Oceans. They also recognize two smaller oceans: the Arctic Ocean, an ocean basin surrounded by continents at the North Pole, and the Southern Ocean, a doughnut-shaped ocean defined by cold waters surrounding the Antarctic landmass. Of course, there are many geographic regions within the oceans: the Mediterranean Sea, the Gulf of Mexico, the Caribbean Sea, the Coral Sea, the Arabian Sea, the Red Sea, the Barents Sea, and numerous others. But these geographic designations disguise an important characteristic of Earth's waters: they don't stay in the same place for long. They

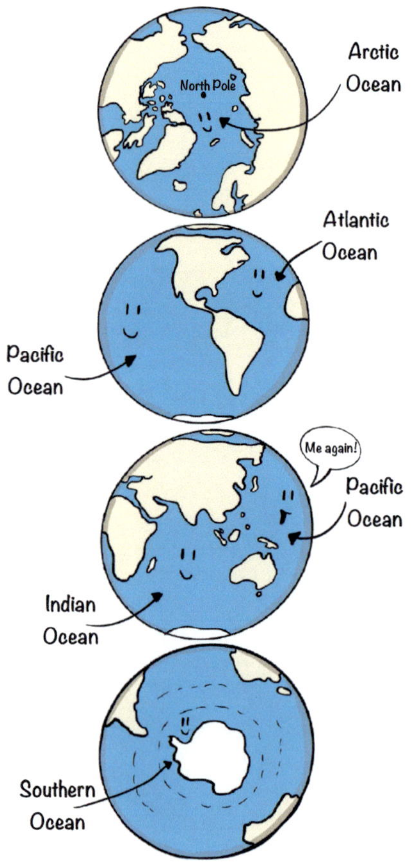

Figure 1.4 The world ocean encompasses the five major geographic oceans.

Credit: Shaw and Chamberlin

Oceanographers, the scientists who study the ocean, define the waters contained within the various ocean basins as one big ocean, the world ocean. This definition permits a view of the world ocean as a system of interdependent parts and processes.

don't even stay in the same ocean basin. Water at one geographic location eventually—over timescales of hundreds of years—finds itself in another geographic location and even back where it began. The waters of the five major oceans flow between the basins and mix to form one ocean—the world ocean.

The idea of a world ocean, as opposed to several independent oceans, occurred to oceanographers at least a century ago, and likely much earlier. Scientific publications expressing the idea of a world ocean are difficult to find prior to 1917, but British author Agnes Giberne (1845–1939), who wrote novels and nonfiction (mostly with Christian themes), clearly states in 1902 her understanding of a world ocean:

> In these days, we know the Ocean as one vast whole. . . . Minor oceans do exist, certainly. We have the Atlantic, North and South; the Pacific, North and South; the Arctic, the Antarctic, and the Indian. Yet for us there is but one great world-wide Ocean, encircling the Earth, every part being in connection with every other part. (Giberne 1902)

Giberne's proclamation notwithstanding, various sources give credit for the term "world ocean" to Russian oceanographer Yuli Shokalsky (1856–1940), who used it in his 1917 monograph *Oceanography*. For English readers, Scottish oceanographer James Johnstone (1870–1932) used "World-Ocean" as the title of the first chapter in *An Introduction to Oceanography*, one of the earliest textbooks on the subject (Johnstone 1923). Johnstone mentions "the great Southern Ocean" and the "three Oceanic 'Gulfs' [that] open out from the Southern Ocean." The idea of the Southern Ocean at the "center" of the three major basins anticipates a

The Southern Ocean exerts a powerful influence on ocean mixing and the world ocean circulation.

modern view of the Southern Ocean as a kind of "steering wheel" for the world ocean circulation (Alfred Wegener Institute 2020).

World ocean thinking appears firmly embedded in oceanographic thinking by the 1930s. American oceanographer Thomas Wayland Vaughn (1870–1952) wrote in *International Aspects of Oceanography* (1937):

> Just when the concept of the unity of all oceans originated is not easy to ascertain. As soon as it was recognized that the cold water in the depths of the oceans had come from Polar regions and that the renewal of the supply of water in those regions had come from other latitudes, the idea of a world ocean was born, and research was directed toward both the circulation within and the exchange of water between the different ocean basins.

South African American oceanographer Athelstan Spilhaus (1911–1998) put it even more succinctly in a paper titled "Maps of the Whole World Ocean" (1942):

> Because it covers more than two-thirds of the earth's surface, a map of the world ocean is essentially a world map. On ordinary world maps, the interruption forming the edges of the map are often placed in the oceans to show the continents to best advantage. If, on the other hand, oceanographic conditions as a whole are to be shown, it is desirable to have the map interrupted within the land masses and the world ocean shown as a unit.

By the 1960s, various authors had published books with "world ocean" in their titles or chapter headings, suggesting common usage of the term by this time (e.g., Carrington 1960; Cromie 1962; Pizer 1967). The first textbook to use the term in its title, *The World Ocean:*

An Introduction to Oceanography (Anikouchine and Sternberg 1973), was the one I used in my freshman year at the University of Washington in 1974. The authors introduce the interconnectivity of the world ocean's waters as the **principle of unity**. And they argue that a focus on memorizing the "names and placements of the continents and ocean . . . is slightly misleading because it suggests that the oceans are separated geographically. From an oceanographer's point of view, the emphasis should be on a world ocean that is completely intercommunicating." Of course, once the idea of a single world ocean is firmly embedded in your thinking, it doesn't hurt to know the names and locations of at least some of the geographic oceans. How else will you know where to go when someone invites you to the Caribbean?

The Wikipedia entry List of Seas (2023) includes more than 370 named oceanic regions. In fact, the world ocean encompasses thousands of informally named bays, bights, estuaries, gulfs, sounds, straits, and similiar bodies of water.

Figure 1.5
There's few experiences more exhilirating than diving the blue waters of Cozumel in the Caribbean Sea.

Credit: istock/Profundo no Mundo

1.2 An Interdisciplinary Science

The scientific study of the world ocean, a field of science called **oceanography**, involves several different scientific disciplines. Generally, oceanographers identify with one of four major subdisciplines:

- **geological oceanography**, the study of geologic processes in the ocean
- **chemical oceanography**, the study of chemical processes in the ocean
- **physical oceanography**, the study of physical processes in the ocean
- **biological oceanography**, the study of biological processes in the ocean

Figure 1.6
Oceanographers generally specialize in one of the four major subdisciplines of oceanography, though familiarity with each can be highly beneficial to research efforts.

Credit: Shaw and Chamberlin

Oceanography is an interdisciplinary science, requiring the expertise of scientists from many different disciplines.

Oceanography also borrows from astronomy, geophysics, meteorology, physiology, ecology, and environmental sciences, to name a few. The integration of many different fields of science into one makes oceanography an **interdisciplinary science**, one that shares and integrates the tools and knowledge of many sciences. Though oceanographers tend to specialize in one discipline, their work requires them to be conversant in other sciences as well. They tend to dabble in multiple subjects, which is why an oceanography course can be so much fun—you learn a little of everything!

Oceanography also involves a host of non-scientist professionals, too, from technicians and programmers to a ship's crew. Technicians provide an onboard repair shop for finicky instruments, programmers maintain the software that communicates with ocean sensors, and the crew navigate the ship, help deploy gear, and cook your meals. (A good rule to follow on board an oceanographic vessel? Be nice to the cook.) Without these support personnel, oceanography would be next to impossible.

Figure 1.7
The people who work in the galley and other ship's crew are as integral to the success of an oceanographic expedition as the technicians and scientists.

Credit: NOAA

Practically speaking, oceanographers seek to understand how human activities affect the ocean's ability to provide food, water, a stable climate, and other ecosystem services.

From a practical standpoint, oceanographers seek to understand the ways that the ocean serves humanity, how it provides what are called **ecosystem services**. Costanza et al. (1997) define ecosystem services as "the ecological characteristics, functions, or processes that directly or indirectly contribute to human well-being." The ocean, for example, provides food, water, materials, waste purification, and a stable climate, among other things. Oceanographers want to know how human activities interfere with the services the ocean provides—how our actions affect the ocean's ability to function properly and provide services to humanity. With more global-scale environmental and social challenges facing us than ever before, the stakes for understanding the ocean have never been higher.

1.3 An Earth Systems View

Humans exist in the web of a vast life support system controlled largely by the ocean. Much of the air that we breathe, the water that we drink, the food that we eat, and the climate in which we live originates from processes that operate within the ocean. While geologic forces ultimately contribute to slow changes in Earth's climate, the ocean—in concert with other systems—dominates processes that affect climate on human timescales (e.g., decades to centuries). Together the various interdependent, interconnected systems that collectively regulate conditions on Earth have come to be known as the **Earth system**.

Loosely defined, a **system** consists of an assemblage of parts and processes working together. System parts include physical components—things composed of matter—while system processes refer to events or activities that result in some outcome. "Working together" implies interactions among those parts and processes—exchanges of energy and matter—and interdependency, meaning the parts and processes depend on each other. Systems may be simple or complex. Like molecules, they may exhibit **emergent properties**, characteristics and behaviors that cannot be observed by studying the individual parts separately. The properties of a water molecule differ markedly from the individual properties of hydrogen (a highly flammable gas) and oxygen (necessary for combustion). With emergent properties, the whole is greater than the sum of its parts.

A great number of systems surround and support you in your daily life. For example, a coffee maker is a fairly simple manmade system. It has parts (stand, filter, heater, etc.), the parts work together (coffee goes in the filter, the filter goes in the filter holder, water pours through the filter), there are regulatory processes for temperature and water flow (adjustable on pricier models), and the system produces a great-tasting end product: a cup of coffee. Yet it would be difficult to describe the taste of a cup of coffee simply by studying the parts of a coffee maker.

Similarly, the human body consists of lots of subsystems: circulatory, nervous, musculoskeletal, urinary, reproductive, and more. But do these individual subsystems really describe you? Of course not. Each of you maintain your own beautiful self, despite the similarities of our bodily systems.

Figure 1.8
The parts and processes of a coffee maker represent a kind of system.

Credit: Shaw and Chamberlin

It was Christmas Eve 1968, but from up there you would never know that a hot war was raging in Vietnam or that a Cold War was dividing Europe. You wouldn't know of the assassinations of Dr. Martin Luther King Jr. or Bobby Kennedy. From that distance, people are invisible, and so are cities, countries, and national boundaries. All that separates us ethnically, culturally, politically, and spiritually is absent from the image. What we see is one fragile planet making its way across the vastness of space.

Dan Rather, *What Unites Us* (2017)

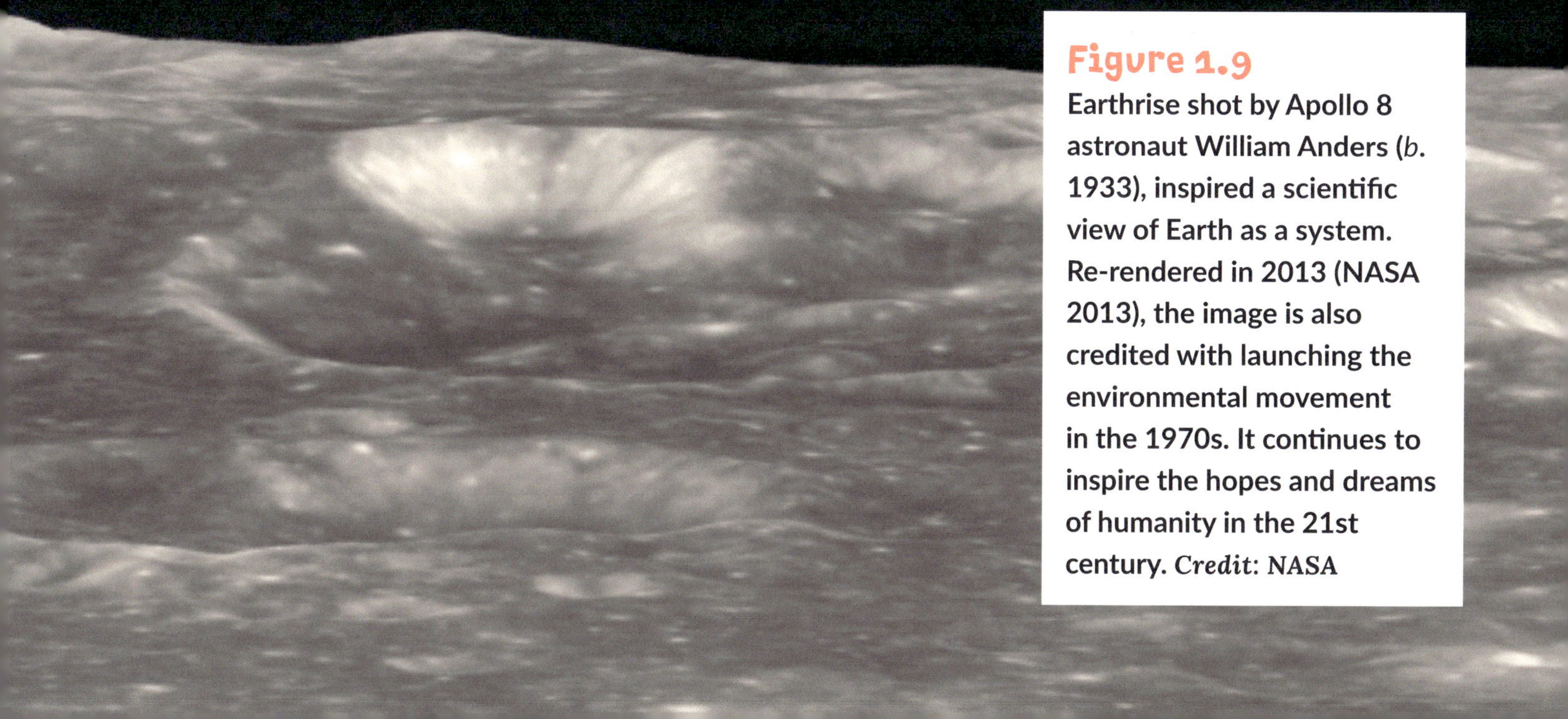

Figure 1.9
Earthrise shot by Apollo 8 astronaut William Anders (*b.* 1933), inspired a scientific view of Earth as a system. Re-rendered in 2013 (NASA 2013), the image is also credited with launching the environmental movement in the 1970s. It continues to inspire the hopes and dreams of humanity in the 21st century. *Credit:* NASA

1.4 Earth's Seven Spheres

Though highly complex, the Earth system operates in much the same way as other systems with which we might be familiar. Within the Earth system, we can identify seven major subsystems, sometimes referred to as spheres (a reference to their names as well as the concept of their operation in a volume of Earth):

- the **heliosphere**, the Sun and solar system
- the **atmosphere**, the gases surrounding Earth
- the **geosphere**, the solid Earth
- the **cryosphere**, the frozen parts of Earth
- the **biosphere**, Earth's living systems
- the **hydrosphere**, Earth's waters
- the **anthroposphere**, the human system

The hydrosphere includes the world ocean and all other waters upon or within Earth's surface, including groundwater. The geosphere shapes the ocean basins and controls exchanges of energy and matter between Earth's crust and the ocean. The atmosphere exchanges energy and matter between the atmosphere and ocean. In the frozen regions of the ocean, the cryosphere, when present, acts as a lid blocking the exchange of energy and matter between the ocean and the atmosphere. All subsystems interact with the Sun, part of the heliosphere, which supplies energy to Earth. Life on Earth constitutes what's known as the biosphere, found from the uppermost reaches of the atmosphere to deep within Earth's crust. One part of the biosphere, humans, have so radically changed our planet that scientists now consider human activities as a separate subsystem, the anthroposphere.

The description of the Earth system as a collective of seven spheres or subsystems is a useful model for understanding complex systems. Scientists call this a **conceptual model**, a sort of mental

Figure 1.10
Earth's seven spheres (from top to bottom, left to right):
1. Heliosphere
2. Atmosphere
3. Geosphere
4. Cryosphere
5. Biosphere
6. Hydrosphere
7. Anthroposphere

Credit: *Shaw and Chamberlin*

picture of how we think something works. Conceptual models guide our thinking and promote insights that can be tested by observation or experimentation. Our model of Earth as a system helps us appreciate the interconnectedness of its geological, chemical, physical, biological, and human processes. Perhaps most importantly, it helps us understand our utter dependency on this system for our survival. As so eloquently put in one of my favorite short documentaries, *The Living Sea*, "We live because the sea lives" (MacGillivray 1995). Conditions established by interactions among the seven spheres make Earth habitable for all living things.

1.5 The World Ocean System

The world ocean may also be viewed as its own system, the world ocean system. The geology, chemistry, physics, and biology of the ocean interact over nearly all scales of space and time. They're all connected! In fact, the study of the world ocean as a system of interdependent parts and processes is what distinguishes oceanography from related disciplines, such as marine biology. Fundamentally, oceanographers aim to understand how the ocean works as a system. That's why it behooves

Figure 1.11 A monk seal checks out a Hawaiian reef. How many of Earth's spheres can you recognize in this photo?

Credit: NOAA CC BY 2.0

Oceanographers view the world ocean as a system of interdependent parts and processes. Its geology, chemistry, physics, and biology are all connected.

oceanographers to be familiar with other subdisciplines. Oceanographic research works best as an interdisciplinary science. We'll visit this idea of an interconnected ocean system throughout the book. Indeed, it's the book's central theme.

1.6 Keeping an Eye on Climate Change

Earth's warming temperatures—**global warming**—and the consequences of that warming for Earth's weather should concern every one of us. Nearly 100 percent of all climate scientists agree that humans are the cause of global warming. Yes, climate varies naturally. But human activities, especially the burning of fossil fuels and the chopping down of forests (i.e., deforestation), have sent Earth's climate beyond its natural variations. Well beyond. Because we're the cause, we have the capacity to be the solution.

Figure 1.12
Earth's temperature is rising because of human activities.

Credit: Shaw and Chamberlin

Figure 1.13
Human-caused climate change has made Earth's weather more extreme.

Credit: istock/mdesigner125

As you might expect, the ocean plays an enormous role in regulating the severity of the changes ahead. But before we go much further in our journey, it's helpful to distinguish what scientists mean when they are talking about weather or climate.

Every day of our lives, we experience the changing conditions of the atmosphere, or what we call **weather**, the instantaneous state of the atmosphere in a particular location. Weather is happening right here, right now. Weather includes several different variables, such as temperature, humidity, clouds, precipitation, and wind. The ocean influences these variables.

As we grow older, we notice patterns in the weather; seasons change, some years feel hotter or colder, some seem wetter or drier. Seasonal, annual, decadal, and even longer timescale patterns in weather define what scientists call **climate**, the long-term statistics of weather conditions at a given location or globally. Climate represents the probability of a particular type of weather at a given time and place. For example, in California we normally expect hot, dry summers followed by cool, wet winters. Of course, any given day may be hot or cold in any season. But climate includes more than just the statistical average of weather conditions (e.g., average temperature or amount of precipitation). It also includes the type of weather event, its frequency and duration, and its intensity (Environmental Protection Agency 2021). Heatwaves, cold snaps, extreme floods, and droughts exhibit patterns over time that make up the statistics of climate.

A sports analogy can help us differentiate between weather and climate. A baseball player may get a hit or strike out in any given at bat. How they perform over a series of at bats defines their batting average. Hitters with a high batting average hit the ball more frequently. They may still strike out, but on average, they're going to hit the ball more times than players with lower batting averages. Climate is like a batting average: it's the weather we expect on average. Weather, however, like a given at bat, is what we get on any given day. A cold day in summer doesn't mean winter has come. It's just the weather for that day.

Figure 1.14
Climate is like a hitter's batting average. With an average of .333, we can expect the player to get a hit every third at bat. But like weather—on any stretch of at bats—they might get several hits or multiple strikeouts in a row.

Credit: istock/IPGGutenbergUKLtd

Figure 1.15
Vast stretches of our coastlines may be underwater in the not-so-distant future, like this statue, Christ of the Abyss, at John Pennekamp Coral Reef State Park in the Florida Keys.

Credit: istock/designimpressario

What worries scientists is that since the mid-1800s, Earth's global climate has been changing faster than normal. Scientists refer to this as **global climate change**—changes in the statistics of weather conditions for the entire planet. Now, climate has been changing throughout Earth's history. During the ice ages, much of North America and Europe was covered in glaciers. At other times, Earth has been so hot that the polar ice caps were completely melted. But the changes that brought these "icehouse" and "hothouse" conditions to Earth happened slowly, over the course of millions of years. The pace of change over the past several decades has no natural counterpart in Earth's past. We hold the key to our climate future. And it begins with our appreciation and understanding of the Earth system.

An ocean-literate person has the scientific knowledge and understanding to advocate sustainable use of the world ocean for the benefit of future generations.

1.7 Ocean Literacy

In 2004, a group of more than 100 scientists, educators, policymakers, and ocean advocates met to discuss the most important ideas about the ocean. Based on those discussions, they published *Ocean Literacy: The Essential Principles of Ocean Sciences* (Ocean Literacy Network 2006) and later *The Essential Principles and Fundamental Concepts of Ocean Sciences for Learners of All Ages* (NOAA 2021). This document outlines seven major principles of ocean literacy, defined as "an understanding of the ocean's influence on people and people's influence on the ocean." *The Essential Principles of Ocean Sciences* provides a road map for understanding our relationship with the ocean. The principles state:

1. The Earth has one big ocean with many features.
2. The ocean and life in the ocean shape the features of the Earth.
3. The ocean is a major influence on weather and climate.
4. The ocean makes Earth habitable.
5. The ocean supports a great diversity of life and ecosystems.
6. The ocean and humans are inextricably interconnected.
7. The ocean is largely unexplored.

This book adopts these principles and aims to develop in you an understanding of them. Like all science education, this effort is part of a global effort to improve **science literacy**, defined by the National Academy of Sciences as "the knowledge and understanding of scientific concepts and processes required for personal decision-making, participation in civic and cultural affairs, and economic productivity" (National Research Council 1996, 2013). Through a study of the ocean, you become not only an ocean-literate person but a science-literate one too. Your ability to "speak science" serves not only your interests but also the interests of the world. An ocean- and science-literate person uses their scientific knowledge and understanding to advocate sustainable use of the world ocean for the benefit of future generations.

1.8 Your Mission

Your mission, ocean explorer, if you decide to accept it, is to develop an understanding of our world ocean, the most important ecosystem on Earth. Through this understanding, I hope that you will dedicate yourself to becoming an ocean-literate person. In traveling these pages, I hope you will see how your life and the world ocean are intimately connected. Ultimately, I hope that what you learn here will inspire you to teach others about the world ocean. Together we can make our planet a more sustainable home for all beings, human and otherwise. As Sengalese conservationist Baba Dioum (*b*. 1937) so eloquently put it, "In the end, we will conserve only what we love; we will love only what we understand and we will understand only what we are taught" (Dioum 1968, as cited by Lotze 2020).

I hope you'll join me. Full speed ahead!

Figure 1.16
Love for the world ocean begins with understanding how it sustains and shapes our lives.

Credit: Shaw and Chamberlin

1.9 Chapter References

Anikouchine, William A., and Richard W. Sternbeg. 1973. *The World Ocean: An Introduction to Oceanography*. Englewood Cliffs, NJ: Prentice Hall. https://archive.org/details/worldoceanintrod00anik

Carrington, Richard. 1960. *A Biography of the Sea: The Story of the World Ocean, Its Life, and Its Influence on Human History*. New York: Basic Books. https://archive.org/details/biographyofseat00carr

Costanza, Robert, Ralph d'Arge, Rudolf de Groot, Stephen Farber, Monica Grasso, Bruce Hannon, Karin Limburg, Shahid Naeem, Robert V. O'Neill, Jose Paruelo, Robert G. Raskin, Paul Sutton, and Marjan van den Belt. 1997. "The Value of the World's Ecosystem Services and Natural Capital." *Nature* 387: 253–260. https://doi.org/10.1038/387253a0

Cromie, William. 1962. *Exploring the Secrets of the Sea*. Englewood Cliffs, NJ: Prentice Hall. https://archive.org/details/exploringsecrets0000crom_i1o3/page/n7/mode/2up

Environmental Protection Agency. 2021. "Climate Change Indicators: Weather and Climate." Updated May 12, 2021. https://www.epa.gov/climate-indicators/weather-climate

Giberne, Agnes. 1902. *The Mighty Deep and What We Know of It*. London: C. Arthur Pearson. https://doi.org/10.5962/bhl.title.18276

Harris, P. T., M. Macmillan-Lawler, J. Rupp, and E. K. Baker. 2014. "Geomorphology of the Oceans." *Marine Geology* 352: 4–24. https://doi.org/10.1016/j.margeo.2014.01.011

Johnstone, James. 1923. *An Introduction to Oceanography*. Boston: Small, Maynard, and Company; Liverpool: The University Press of Liverpool. https://doi.org/10.5962/bhl.title.25450

Lotze, Heiki K. 2020. "Combining Love and Knowledge to Heal the Ocean." *Ethics in Science and Environmental Politics* 20: 33–39. https://doi.org/10.3354/esep00193

MacGillivray, Greg, director. 1995. *The Living Sea*. Laguna Beach, CA: MacGillivray Freeman Films. Film, 40 minutes. https://www.imdb.com/title/tt0113676/

NASA. 2018. "Earthrise 1: Historic Image Remastered." NASA Science, December 20, 2013. https://earthobservatory.nasa.gov/images/82693/earthrise-revisited

National Research Council. 1996. *National Science Education Standards*. Washington, DC: The National Academies Press. https://doi.org/10.17226/4962

———. 2013. *Next Generation Science Standards: For States, by States*. Washington, DC: The National Academies Press. https://doi.org/10.17226/18290

NOAA. 2021. *Ocean Literacy: The Essential Principles and Fundamental Concepts of Ocean Sciences for Learners of All Ages*. Washington, DC: NOAA. https://www.marine-ed.org/ocean-literacy/overview

Ocean Literacy Network. 2006. *Ocean Literacy: The Essential Principles of Ocean Sciences, K–12*. Washington, DC: The National Geographic Society. https://www.aquariumofpacific.org/downloads/ed_SevenEssentialPrinciples.pdf

Pizer, Vernon. 1967. *The World Ocean: Man's Last Frontier*. Cleveland and New York: World Publishing. https://www.google.com/books/edition/The_World_Ocean/O74z2lnoMZwC

Rather, Dan. 2017. *What Unites Us: Reflections on Patriotism*. Chapel Hill: Algonquin Books. https://www.workman.com/products/what-unites-us/paperback

Sagan, Carl, and Ann Druyan. 2011. *Pale Blue Dot: A Vision of the Human Future in Space*. New York: Ballantine Books. https://www.penguinrandomhouse.com/books/159735/pale-blue-dot-by-carl-sagan/

Shokalsky, Yu. 1917. *Oceanography*. Petrograd.

Spilhaus, Athelstan F. 1942. "Maps of the Whole World Ocean." *Geographical Review* 32(3): 431. https://doi.org/10.2307/210385.

Vaughan, Thomas Wayland. 1937. *International Aspects of Oceanography: Oceanographic Data and Provisions for Oceanographic Research*. Washington, DC: National Academy of Sciences. https://doi.org/10.5962/bhl.title.16994

Wikipedia. 2023. "List of Seas." Wikipedia, March 4, 2023. https://en.wikipedia.org/wiki/List_of_seas

Figure 1.17
All creatures—large and small—play a vital role in the world ocean system.

Credit: Shaw and Chamberlin

Nothing escapes their eye.
Artwork by Martha Rich.

Team Oceanography

2

Figure 2.1
Oceanographers are scientific explorers of the natural wonders of the world ocean.

Credit: istock/ifish

Our understanding of how the world ocean works as a system emerges from the **scientific method**, the systematic and self-correcting set of practices that scientists use to discover knowledge about the natural world. But while textbooks often present the scientific method as a set of rigid rules that scientists must follow, I'm here to tell you that science is messy—occasionally very messy—with twists and turns as dramatic and spine-tingling as an Agatha Christie novel. As you'll see, the scientific "method" is more of a scientific "process," a back-and-forth, collaborative, social—and did I say messy?—web of activities carried out by a community of scientists on a quest to understand how nature works.

The Hollywood model of the scientific method goes something like this: A scientist—usually a man in a white lab coat working in a dark basement filled with bubbling test tubes—poses a question, creates a hypothesis, tests the hypothesis, gets a result, forms a conclusion, communicates the results, and wins a Nobel Prize. The crowd roars. But science rarely, if ever, works this way.

Scientific research proceeds in fits and starts, maybe even two steps forward, one backward, and a couple sideways. The standard sequence may be taken out of order. An initial approach may prove to be a dead end. Early results may be shared to get feedback. Methods may be refined. Hypotheses may be revised, and revised again, as new observations and measurements lead to a deeper understanding of a problem. All kinds of things, theoretical, methodological, statistical, and conceptual, may interrupt the smooth sequence of textbook science. Instead, scientists multitask and work with other scientists and skillfully jump between different activities to ensure that by the end of a particular study, the data they collect, the methods they use, and the interpretations they offer meet rigorous standards and pass muster with their fellow scientists, a key step called peer review.

This modern approach has been called the **activity model of the scientific method** (Harwood 2004). The activity model emphasizes the social character of science, which proceeds in small, careful steps with numerous refinements and do-overs. Preliminary results are scrutinized to refine methods and make adjustments prior to full-scale implementation. Scientists may share data in the cloud or an online database to gain feedback from other scientists for improving their approach or to support scientists working on similar problems. An exchange of ideas and new findings happens via telephone, email, social media, and at any of the regular gatherings of professional organizations, scientific conferences, or virtual meetings. In this way, science in the 21st century takes on many of the aspects of a thriving, dynamic, and exciting entrepreneurial enterprise.

Oceanography, perhaps more than any other discipline, embodies the activity model of the scientific method. If you're looking for a scientific career that brings you into contact with people from all walks of life, you'll likely enjoy working in ocean science. We call it team oceanography. Come join us!

Figure 2.2
Ocean science requires the efforts of geologists, chemists, physicists, biologists, ecologists, and more.

Credit: Shaw and Chamberlin

Figure 2.3
Teamwork makes mud-sifting a lot more fun (and faster) in the frigid Arctic.

Credit: **NOAA/*Katrin Iken***

2.1 Channel Your Inner Scientist

To understand the scientific process, it's helpful to learn how to think like a scientist—a real scientist, not the fact-quoting one you learned about in secondary school or the nerdy one on TV (although I do love Sheldon). In fact, your natural "inner child" instincts are those of a scientist.

Anyone who spends time with young children appreciates their innate curiosity, their desire to know. Not long after children learn to speak, they learn to ask "what" and "why." "What's that?" they may say, pointing to a squirrel in a tree. "Why did the cat bite me?" they might cry, learning to never again pull the cat by its tail. "Why do stars fall down from the sky?" as they point to the trail of a shooting star.

2.2 Investigate the Natural World

Science starts with curiosity. To see the world as scientists do requires a childlike wonder to see and hear and smell and touch and taste the world in high definition. In the lab or in the field, a scientist's senses are

Science begins with a childlike curiosity about the ntural world.

on alert. Every detail matters. Like the Bene Gesserit of the 1965 novel *Dune*—written by Frank Herbert (1920–1986) and considered the best-selling science fiction novel of all time—scientists are "trained in the minutiae of observation." Only through repeated observations of the same organism, behavior, event, or occurrence can scientists establish a set of facts upon which experiments or further observations can be based.

Phenomena that can be observed and measured consistently and repeatedly lend themselves to **scientific inquiry**, the multifaceted process of investigating the natural world. Things that cannot be observed or measured, or that occur once or rarely, do not lend themselves to scientific inquiry. Such phenomena fall outside the domain of science. Scientific inquiry encompasses a diverse set of tools and approaches. To the extent that the researchers carrying out the inquiry report their methods and results truthfully, and as long as they remain willing to modify their conclusions based on their results, then these efforts may be considered legitimate science.

Figure 2.4 Curiosity about the natural world begins at a very young age.

Credit: istock/FluxFactory

2.3 Ask a Good Question

Science aims to ask questions about the universe and everything in it, including us. In my classroom and online teaching experience, it's a rare student who asks a question about a topic or concept being covered in the course. Perhaps students fear that they will appear unknowledgeable and their instructor or their classmates will judge them. Maybe it's because nobody likes a smarty-pants (e.g., Rentzsch and Schröder-Abé 2014). Or maybe students just aren't curious about the topics on the syllabus—the saddest of all possibilities.

Whatever the reason, I like to think that many students don't realize that asking questions is a good thing. Journalists and scientists learn very early in their careers that asking great questions reaps great benefits. Experienced journalists and scientists can ask questions that other folks never think of. Just like surfing Bonzai Pipeline takes practice (and a bit of nerve), asking good questions takes practice (and a bit of confidence). Like the big wave that got away, an unasked question is an opportunity missed.

In science, however, we differentiate between scientific and nonscientific questions. A **scientific question** concerns the natural and

Figure 2.5
Good science requires careful observation.

Credit: istock/KirsanovV

observable world. As outlined by the National Research Council (NRC 2002), a scientific question is one that "can be investigated empirically," that is, by using our senses to gather evidence. Questions like why is the ocean blue, why do waves break at the shore, and how do sperm whales hold their breath so long can be answered using science. Questions about if God exists or whether humans have untapped mental powers aren't scientific. They don't lend themselves to observation (at least not at this time).

A scientific question typically builds upon previous research and scientific theory; it "links research to relevant theory" (NRC 2002). A newly posed scientific question contributes to a **chain of reasoning**, a logical extension of previous work. It's not enough to simply work on a question in a field of study in which others are working. A scientist's work requires an intimate knowledge of what has been published in that field and what questions remain unanswered. A scientist's questions pick up where others left off or take the next step in a series designed to figure

Figure 2.6 Science starts with testable questions based on observations of the natural world.

Credit: Shaw and Chamberlin

Scientific questions build from one to the other. Answers to one question generate new questions and new lines of inquiry. The history of science reads like an ancestral tree of ideas, some with lineages extending centuries; others, gone in a generation.

something out. A 12th-century quote popularized by Sir Isaac Newton (1643–1727) sums it up nicely: "If I have seen further, it is by standing on the shoulders of giants" (Newton 1675). The discoveries of modern science owe their existence to the discoveries of scientists in previous times.

Scientific questions must be general enough to apply in different settings. A study that focuses on one event in one place at one time will be difficult to replicate and generalize. Research on human behavior in social sciences and education often faces problems of replicability and generalization. Replicability stands at the heart of scientific progress. If a scientist's results cannot be replicated, then those results will be questioned and potentially dismissed.

Figure 2.7
Multiple observations ensure reliable and replicable results. Hundreds of measurements of ocean temperature may be taken on a single cruise.

Credit: NOAA

2.4 Propose an Idea or Model

Collecting observations and forming questions naturally lead to consideration of possible answers. Humans relish in coming up with ideas about why something happened. Because of the way our brains work, we try to fit observations into mental frameworks of what we already know. These frameworks become conceptual models (introduced in Chapter 1), idealized representations of the world around us.

Conceptual models often take the form of illustrations, diagrams, and even animations. Descriptions and drawings of imaginary things like dragons, mermaids, and extraterrestrials represent someone's conceptual model of these creatures. Ideas also take the form of illustrations or graphics, primarily as a means of refining those ideas and explaining them to others. In fact, most illustrations that you see in textbooks represent conceptual models. An illustration of the major ocean currents and a diagram of the interactions of species in an oceanic food web serve as examples of conceptual models of how a part of the ocean works.

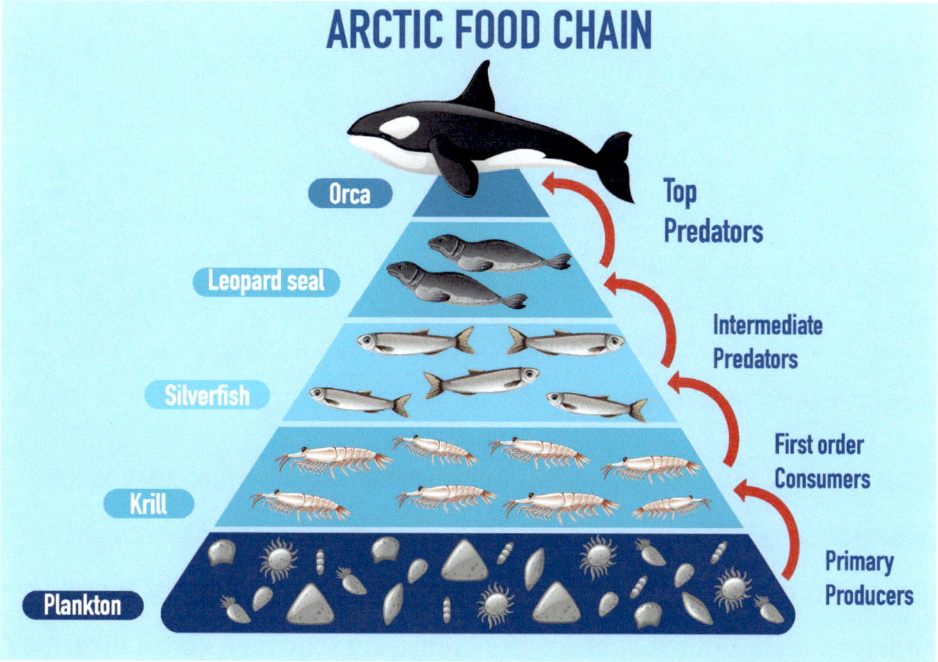

Credit: istock/blueringmedia

Figure 2.8
Conceptual model of an Arctic food chain illustrating the transfer of foodstuffs from those that produce it (base of pyramid) to those that consume it (top of pyramid).

Scientists, like everyone else, employ other kinds of models to help them understand and test hypotheses about the natural world. As a child, you likely played with toys. Toys are a kind of **physical model**, a scaled-down, three-dimensional, semirealistic representation of an object or part of nature. A Matchbox car is a physical model of a real car. A Cabbage Patch doll is a physical model of a human baby. Lincoln Logs represent real logs; Legos represent construction materials. These physical models allow kids to experience through pretend play what it's like to drive a real car, care for a real baby, or build a house. They provide a sense of the excitement and, hopefully, the responsibilities involved with an actual car, baby, or house.

Figure 2.9
People can be physical models. A photographer's physical models may represent a style of fashion or particular lifestyle.

Credit: istock/g-stockstudio

A third kind of model uses mathematics to formulate and test ideas. **Mathematical models** represent simple or complex expressions used to describe or simulate how something works. We usually think of them as **quantitative**—giving numerical results—but they may also be **qualitative**, helping scientists describe properties or relationships between

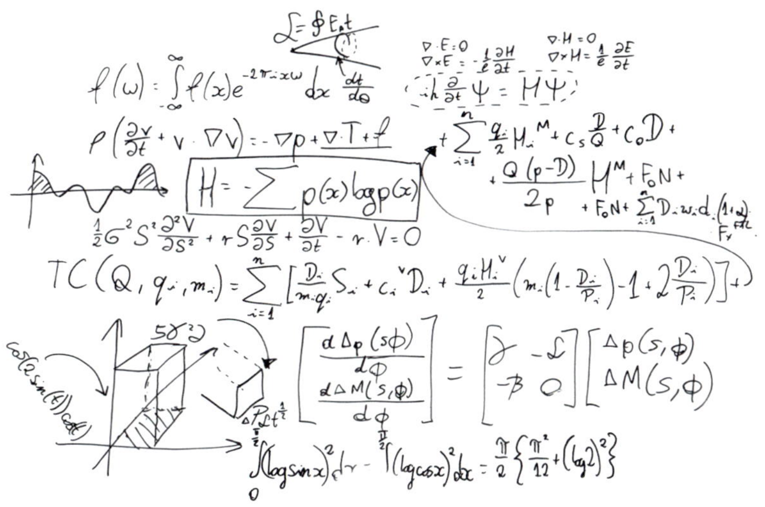

Figure 2.10
Much as some of us may dislike math, it's an indispenable tool for making predictions and helping scientists understand how nature works.

Credit: istock/NiseriN

variables (e.g., Dambacher et al. 2015). Mathematical models underpin the stock market, retirement planning, business development, economics, and a whole host of other human activities.

2.5 Test Your Hypotheses

Models form the cornerstone of human thinking, but a model, any model—conceptual, physical, or mathematical—is only as good as the facts that support it. To determine whether a model accurately describes or predicts a natural event, scientists carry out what may be the most important part of their work: creating testable hypotheses.

A **hypothesis** is a carefully worded, conditional, and testable explanation of how nature works. You'll often hear people call them educated guesses, but there's little guesswork involved. Scientists typically frame hypotheses within the context of known explanations. Because of previous observations and experiments, they often have a pretty good idea of what's happening—or at least they think they do—and so hypotheses represent the next step in a progression of observations and experiments designed to figure something out. They aren't random and they aren't long lists that include every possible explanation. They are focused, narrowly interpreted, tentative statements that can be tested

Science works by disproving hypotheses, a process called falsification. When observations or experiments fail to reject a hypothesis, only then may it be considered a valid statement of the phenomenon under investigation.

using carefully planned observations and carefully designed experiments. For example, you might state the hypothesis, "Whales jump out of the water to remove barnacles from their skin." You could test this by observing whales.

One often overlooked and misunderstood characteristic of hypotheses is how they are worded. Scientists pose hypotheses in a manner that allows them to be rejected by observations and experiments. It may seem a curious way to go about something, but science generally attempts to disprove possibilities—eliminate hypotheses—rather than prove them. Scientists call this method **falsification**, the process of disproving a hypothesis. While some argue that falsification can be limiting in certain fields, especially branches of modern particle and theoretical physics (e.g., Carroll 2014), most scientists view it as a cornerstone of the scientific method. Our test of the whale-jumping hypothesis would be rejected (i.e., falsified) if we observed whales without barnacles jumping or if the act of jumping had no effect on whale-attached barnacles. See how that works?

Figure 2.11 Models help scientists generate testable hypotheses.

Credit: Shaw and Chamberlin

The process of eliminating competing hypotheses was developed by Austrian-British science philosopher Karl Popper

(1902–1994; Popper 2005). It belongs to a broader approach called **deductive reasoning**. Deduction operates as a kind of top-down logic: If A is true, and B belongs to A, then B is also true. For example, if all mammals breathe air, and a dolphin is a mammal, then dolphins breathe air.

While most scientists use deductive reasoning, the opposite form, **inductive reasoning**, also plays a part in science. Inductive reasoning operates as a bottom-up approach to figuring something out. Specific observations lead to general principles. For example, if dolphins use tools, and sea lions use tools, and both dolphins and sea lions are mammals, then, inductively thinking, all mammals use tools. Induction helps in the process of generating appropriate scientific questions and hypotheses from observations or preliminary experiments. Many scientists consider this the most interesting part of their research because it draws upon the creative mind. Science requires a person to imagine all the possibilities. For that reason, scientists tend to be highly creative people, despite their characterization as boring nerds.

A third form of reasoning, **abductive reasoning**, plays a larger role outside of science, where explanations or conclusions are needed to fix something, make a diagnosis, or render a judgment. Abductive

Figure 2.12 Car mechanics often use abductive reasoning to diagnose an engine problem.

Credit: istock/Sanja85

Figure 2.13 Oceanographers propose different hypotheses for why male humpback whales leap out of the water. They may be trying to impress potential mates or ward off competitors.

reasoning is best explained as the quickest way to the most logical and usually simplest explanation, even though facts and observations may be incomplete. Car mechanics, doctors, and jurors often use abductive reasoning to complete a task because they are presented with limited data. Scientists often use abductive reasoning to generate hypotheses. It's considered the more creative approach to scientific inquiry.

Credit: Shaw and Chamberlin

Figure 2.14 A well-formulated hypothesis provides clear and testable predictions that are easily falsified by observation or experimentation.

Hypothesis making through inductive and abductive reasoning and hypothesis falsification through deductive reasoning make up what's known as the **hypothetico-deductive approach**, also known as the hypothetico-deductive model or simply the H-D method. It relies on the process of

generating a hypothesis that can be falsified by a set of observations or experiments and then carrying out those observations or experiments to reject or support the hypothesis.

The H-D method remains a mainstay of science because it aspires to objectivity. Scientists go to great lengths to be impartial in their work from beginning to end. They don't always get it right, but the self-correcting nature of science and the willingness of scientists to admit their errors make the H-D method the best thing going. To quote American philosopher and education reformer John Dewey (1859–1952):

> Science represents the safeguard of the [human] race against . . . natural propensities and the evils that flow from them. It consists of the special appliances and methods which the race has slowly worked out in order to conduct reflection under conditions whereby its procedures and results are tested. . . . Without initiation into the scientific spirit one is not in possession of the best tools which humanity has so far devised. (Dewey 1916)

Modern civilization arose through application of science to fight disease, improve food production, and develop new energy resources. It helped establish comfortable dwellings and workplaces. It revolutionized how we communicate. Science remains the most promising and hopeful approach for solving the world's problems.

2.6 Conduct an Investigation

All scientific investigations take time, patience, and funding. But ocean science presents special challenges. At least eight broad and overlapping approaches characterize the discipline of oceanography:

- **laboratory investigation**, usually carried out in a laboratory or special facility outfitted with instrumentation, artificial habitats, and other tools for conducting controlled experiments and observations
- **oceangoing (or field) investigation**, typically carried out in the ocean, including along the shore, using snorkel or scuba, or on a small vessel, large ship, submersible, or other manned vehicle
- **robotic investigation**, using remotely operated vehicles, autonomous vehicles, drifters, gliders, and similar craft
- **time-series investigation**, using instruments deployed on moorings, autonomous seafloor instruments, and cabled ocean observatories
- **animal-tagging investigation**, in which oceanographers attach instrument packages to marine mammals (e.g., elephant seals) or large fishes (e.g., great white sharks)
- **aerial investigation** (confined to Earth's atmosphere), including instruments flown on drones, tethered balloons, and manned aircraft
- **spaceborne investigation**, mostly carried out by instruments aboard satellites but occasionally conducted by instruments and personnel aboard spacecraft and space stations
- **computer modeling investigation**, ranging from studies done on personal computers to those requiring supercomputers

An oceanographic research program may include any or all of the above approaches. Investigations may be carried out by a single investigator or multiple investigators. They may be supported by undergraduate and graduate research assistants, postdoctoral researchers, lab and field technicians, and personnel from private industry. An investigator may work with colleagues at their own institution or collaborate with colleagues, students, technicians from multiple institutions. The largest oceanographic programs involve scientists from across the globe cooperating on different aspects of a scientific problem and sharing the costs of personnel, equipment, travel, and ships. It's not uncommon to find people from different countries spending several weeks at sea together, working through language barriers, getting used to different customs, and attempting to enjoy what are often unfamiliar foods (like fiskeboller). It works (most of the time) because everyone depends on each other for assistance and safety and is focused on a common goal.

Credit: Shaw and Chamberlin

Figure 2.15
A field investigation may involve observation or experimentation.

Perhaps more than any other science, oceanography depends on teams of diverse scientists, technicians, and support personnel to achieve its goals. Oceanography has rightfully been called a "team science" (Rudzin et al. 2022).

As a real-life example, an international expedition involving hundreds of researchers from 19 countries began a yearlong mission aboard the icebreaker *Polarstern* in the Arctic Ocean in September 2019. Dubbed the Multidisciplinary drifting Observatory for the Study of Arctic Climate, or MOSAiC, the expedition froze their ship into the ice to conduct investigations designed to shed new light on climate change. MOSAiC, the largest shipboard polar expedition ever mounted, recalls the historic journey of Norwegian oceanographer Fridtjof Nansen (1861–1930). Nansen froze his round-bottom wooden sailing ship, *Fram*, into the Arctic ice from 1893 to 1896. (He's also the only oceanographer to win a Nobel Prize, albeit for his humanitarian work.) Oceanographers participating in the MOSAiC mission—which ended in October 2020—were able to take measurements and perform experiments at polar latitudes over all four seasons, including Arctic winter. Though larger than most, MOSAiC serves as just one example of dozens of similar oceanographic investigations carried out in coastal waters and across the world ocean annually.

Figure 2.16

A new team of scientists, technicians, and crew board the German icebreaker *Polarstern*, frozen into the Arctic sea ice as part of the MOSAiC Expedition.

Credit: *Alfred-Wegener-Institut/Michael Gutsche* CC BY 4.0

2.7 Analyze Your Results

Data analysis represents the worst of times and the best of times for scientists. Entering, organizing, reducing, summarizing, tabulating, graphing, and analyzing scientific observations and data—in some cases, millions upon millions of data points—can be time consuming, frustrating, tedious, and a downright pain in the caboose. On the other hand, as the averages emerge, as the trends become clear, as the relationships make themselves known, and as something that no one has previously witnessed or discovered sees the light of day, data analysis can be one of the most exhilarating experiences in a scientist's life. After all, the data tell the story. They refute or confirm the hypotheses and ideas. They tell the scientists whether they're on the right track or whether a particular set of experiments is an utter failure. Data analysis is often the make-or-break moment in a scientific study. Occasionally, a scientist's work results in something so extraordinary that it changes the world. Our modern, industrialized civilization—including the smartphone you're checking as you read this—represents an accumulation of these moments in science.

Without question, the tools, methods, and approaches of data analysis have been transformed by the development of technology, notably computing. Scientists have been using computers since the 1950s, but computers' speed and ease of use have changed dramatically in the past 20 years. As one oceanographer put it, computers did for the 20th century what adoption of Arabic numerals did for the 17th century: they transformed how we think (Warren 2006).

Figure 2.17
Oftentimes, an initial hypothesis may be disproven when the data are carefully analyzed.

Credit: Shaw and Chamberlin

Figure 2.18

Polar oceanographic research expeditions like MOSAiC often involve excursions onto the sea ice, a welcome break from the rolling (and often noisy) ship.

Figure 2.19
Ferryboats, such as those that cross Puget Sound in Washington, require up-to-date information on weather and ocean conditions to operate safely.

Credit: istock/suwaneeredhead

Three examples illustrate Warren's point: numerical models, big data, and artificial intelligence.

In recent decades **numerical models**, also known as mathematical or computer models, have evolved alongside theory and observation into what oceanographers call the third element of oceanographic research. Models generate data by solving sets of equations that represent—in the scientist's mind—how a system works. When the model's predictions (or forecasts) are compared to observations from nature, scientists can determine how well they do or don't understand a system. When the model outputs fail to match observations, scientists know that some part of their model needs revision. This stepwise process—generating model predictions, comparing them to real-world observations, and revising the model—helps scientists improve their understanding of the natural world.

Models serve a practical function too. A branch of applied ocean science known as **operational oceanography** serves as a kind of ocean weather forecasting service. Its models produce near-real-time output called **nowcasts**. To create nowcasts, oceanographers use observational data from various near-real-time electronic streams, such as buoys, floats, satellites, and even ships. Operational oceanography serves commercial marine interests (fishing and shipping, for example), the

Figure 2.20 Oceanographers rely on computers for data analysis, modeling, and visualization of the enormous data sets generated by oceanographic research.

Credit: NOAA

military (which depends on up-to-date knowledge of the ocean to conduct its operations), and recreational interests (like boaters and surfers).

Increasingly, oceanographers are turning to computer scientists for help managing large oceanographic data sets. **Big data** refers to an emerging subdiscipline of computer science that tackles the challenges of analyzing data sets whose size or complexity exceeds the capabilities of traditional software and computers. Hey et al. (2009) refer to this "data-intensive computing" as the **fourth paradigm**, the use of computers to explore and mine big data for information. In a sense, the fourth paradigm emerges from the ability of computers to think. They detect patterns that we cannot. Big data scientists emphasize five key properties of big data: volume (lots of it), variety (many different sources), velocity (coming at you fast), veracity (ensuring the data are real and free of errors), and value (what you can learn; e.g., De Mauro et al. 2016). The terminology has largely developed in response to the proliferation of electronic devices such as smartphones, watches, tablets, virtual assistants, and similar devices. Capable of tracking, storing, and archiving (via the cloud) practically everything you do on your device, the resultant data stream (popularly known as digital exhaust) generates mind-blowing quantities of information.

Oceanographers have long faced the challenges of big data, **marine big data**, as they refer to it (e.g., Huang et al. 2015, 2019). They note that American oceanographer Matthew Maury's (1806–1873) compilation of wind and current data from ships' logs contained 1.2 million data points. Results from the world's first global oceanographic voyage, the *Challenger* expedition (1872–1876), spanned 50 volumes and more than 29,500 pages (Murray 1895). More recently, large, multi-institutional oceanographic expeditions churn out data by the petabyte. That's a thousand times a terabyte, the typical storage capacity of the current generation of desktop computers. A single Earth-observing satellite, the *Aquarius*, logs more data in two months than the first 125 years of ship and buoy data gathering (Huang et al. 2015). Developing the architecture and algorithms to collect, verify, store, process, integrate, and analyze such massive volumes will require new ways of thinking and a new generation of marine data science techies. Perhaps you will be among them.

One possible solution for extracting meaningful information from big data lies in **artificial intelligence**, or AI, a broad category of data science aimed at building computer and software systems that mimic human intention, intelligence, and adaptability (e.g., West and Allen 2018). While application of AI-based data analysis in the marine sciences has been slow to develop, the field holds great promise for extracting greater value from big data. Working side by side, humans and machines engaged in ocean science research may accelerate our knowledge and understanding of the world ocean as never before.

The amount of oceanographic data collected in modern times is nothing short of staggering. A single satellite collects more data in a few months than were collected in the first 125 years of oceanographic research.

Figure 2.21 Virtual reality holds promise for helping oceanographers visualize big data sets and see the ocean as never before.

Credit: istock/Thinkhubstudio

2.8 Form Your Conclusions

The grand finale of the scientific method—forming your conclusions—proves to be the most challenging for many scientists. This is where the rubber meets the road, so to speak. A scientist must now make sense of their data. They must extract meaning and significance. They must use the evidence of their observations, measurements, and experiments to tell a story. And they must tell that story in a way that other scientists and even the public can understand. Most often, this story forms the basis of a **scientific discussion**, the presentation in oral or written form of the conclusions of a particular scientific study. Though forming conclusions differs depending on the nature of the research, in general, four steps are crucial (e.g., Heard 2016):

1. How do the results shed light on (i.e., answer) your original research questions?
2. What are the strengths and weaknesses of your results in answering your research questions?
3. How do your conclusions build upon, extend, or contradict previous work on this topic?
4. What questions and considerations arise for future research on this topic?

Figure 2.22
Making sense of your data and communicating their meaning in the context of previous research form the basis of a good scientific discussion.

Credit: Shaw and Chamberlin

Here's where science most often tests the modesty of a scientist. After all, if you designed and carried out the research, you're naturally (and forgivably) inclined to think it's brilliant! Yet a scientist's interpretation of their results may be subject to **bias**, a less-than-objective, partial, and sometimes distorted view of scientific evidence. Scientific bias almost certainly stems from a worldview formed as a result of years or decades of experience in the field. There's nothing intentional or malevolent about it. Scientists are human, after all, but they must always remain aware of the inherent biases in their work. As Milroy (2016) puts it:

> Regardless of the strength of your hypothesis or the veracity of your data, as human beings we are always prone to making errors in our analyses. In the context of science, this almost always occurs because we have some preconceived notion of how our experiments should turn out, and if our data seem to support our original hypothesis, it is difficult to step back from all of your hard work and look at the results with a truly cynical eye. Results can be easily misinterpreted unless you are very careful to devise an experiment where the results can only be construed in one way.

The scientific method embodies a set of checks and balances that guard against this very thing, a system called **peer review**, which may be

Figure 2.23 Fullerton College student Jonathan Yeme (at right) presents the results of his aquaponics research at a symposium sponsored by Cal State Fullerton. Poster presentations are a good way to get feedback on your work.

Credit: Chamberlin

defined as the feedback on a scientist's work by other scientists or experts. A scientist may meet with or Zoom with a trusted colleague and ask them what they think, or present preliminary conclusions to gain feedback. Scientific papers undergo scrutiny by anonymous reviewers, a formal peer review process that further ensures the validity of the data and the veracity of a scientist's conclusions. Of course, errors, misstatements, and, on rare occasions, deliberate misrepresentation of results may occur. But science rests on the accumulation of evidence from multiple studies and multiple scientists. The self-correcting nature of science is what makes it one of the most valuable tools ever invented by humans. Scientists don't always get it right the first time, but eventually, through the scrutiny of the ages, after decades to centuries, the truth emerges. Truths become well-established scientific theories, and the extent of our knowledge and understanding progresses.

Undergraduate research is a great way to gain experience in your discipline, even non-scientific ones.

2.9 Communicate Your Science

In the 21st century, communication of scientific results takes many forms, including:

- peer-reviewed scientific journals
- talks and poster presentations at scientific conferences
- workshop or symposium compilations
- popular magazines (e.g., *National Geographic*, *Scientific American*)
- science books (of all varieties, including textbooks)
- public social media (especially Twitter, but also Facebook and Instagram)
- professional and scientific social media (such as LinkedIn, ResearchGate, Academia)
- videos (e.g., YouTube, Vimeo, TikTok)
- live appearances (e.g., Zoom)
- blogs, podcasts, websites, email, and more!

Unfortunately, most scientists are trained only to present papers and talks. That's fine if you're communicating solely with scientists in your own discipline. But nearly all forms of science these days—and the agencies that fund the work—require communication with people outside the discipline, including scientists in other fields.

In its 2017 report, *Communicating Science Effectively: A Research Agenda*, the National Academy of Sciences, Engineering, and Medicine (NASEM) defines **science communication**, popularly known as SciComm, as "the exchange of information and viewpoints about science to achieve a goal." A science communicator may strive to develop "a greater understanding of science and the scientific method" or to "help people understand the science relevant to a decision." NASEM acknowledges that:

> The methods scientists use to understand the world are unlike the ways people typically think on a day-to-day basis. The results of science also can be insufficient, ambiguous, or uncertain, and scientific conclusions can change over time as new findings emerge. These inherent characteristics of science can create barriers to communication and understanding.

You can't really blame the scientists. American biologist and book author Paul Ehrlich (*b*. 1932) famously opined that "If what [a scientist] does is comprehensible to the general public, it means he's not a good scientist. That's what I thought. I was wrong" (from Goodell 1977). Indeed, the importance of communicating science to the public is greater than ever before. Public interest in climate change, vaccines, environmental pollution, genetically modified foods, fracking, and a host of similar socioscientific issues underscores the need to bridge the gap between scientists and the public. As Fischhoff (2013) puts it, "Effective SciComm informs people about the benefits, risks, and other costs of their decisions, thereby allowing them to make sound choices." Increasingly, educators view SciComm training as a priority in the education of future scientists (e.g., Brownell et al. 2013; Montgomery et al. 2022).

Even if you're pursuing a career outside science, you play a role in SciComm. When you tell family or friends about your science courses, you represent something about science. Your experience in a science class (and reading a science textbook!) shapes your opinions, good, bad, or otherwise. And those opinions are expressed every time you post something (hopefully, good) about science.

Credit: Shaw and Chamberlin

Figure 2.24 Effective SciComm builds public trust in scientists and science.

A common refrain I hear from students is "I'm not good at science." Of course not! You're probably not good at fixing spaceships, either. These things take years of training and experience. But having a sense of where scientists are coming from, learning how to decipher their statements, and understanding that behind every statement is a body of evidence may be more valuable to the world than just knowing a bunch of science facts. If the end result of a science course is that you understand and appreciate how science works—and have a positive feeling about the work that scientists do—then you deserve an A in my book!

Figure 2.25
The future of ocean science rests on our ability to inspire young people to pursue their passion for the world ocean.

Credit: istock/FatCamera

2.10 Chapter References

Alfred Wegener Institute. "MOSAiC." Accessed May 27, 2022. https://mosaic-expedition.org/

Brownell, Sara E., Jordan V. Price, and Lawrence Steinman. 2013. "Science Communication to the General Public: Why We Need to Teach Undergraduate and Graduate Students This Skill as Part of Their Formal Scientific Training." *Journal of Undergraduate Neuroscience Education* 12(1): E6–E10. https://pubmed.ncbi.nlm.nih.gov/24319399

Carroll, Sean. 2014. *Falsifiability*. Contribution to *What Scientific Idea Is Ready for Retirement*? Edge.org. https://www.edge.org/response-detail/25322

Dambacher, Jeffrey M., Peter C. Rothlisberg, and Neil R. Loneragan. 2015. "Qualitative Mathematical Models to Support Ecosystem-Based Management of Australia's Northern Prawn Fishery." *Ecological Applications* 25(1): 278–298. https://doi.org/10.1890/13-2030.1

De Mauro, Andrea, Marco Greco, and Michele Grimaldi. 2016. "A Formal Definition of Big Data Based on Its Essential Features." *Library Review* 65(3): 122–135. https://doi.org/10.1108/LR-06-2015-0061

Dewey, J. 1916. *Democracy and Education: An Introduction to the Philosophy of Education*. New York: Macmillan. https://www.google.com/books/edition/Democracy_and_Education/8P0AAAAAYAAJ?hl=en

Fischhoff, Baruch. 2013. "The Sciences of Science Communication." *Proceedings of the National Academy of Sciences* 110: 14033–14039. https://doi.org/10.1073/pnas.1213273110

Goodell, Rae. 1977. "The Visible Scientists." *The Sciences* 17(1): 6–9. https://doi.org/10.1002/j.2326-1951.1977.tb01494.x

Harwood, William. 2004. "A New Model for Inquiry: Is the Scientific Method Dead?" *Journal of College Science Teaching* 33(7): 29–33. https://www.jstor.org/stable/10.2307/26491315

Heard, Stephen B. 2016. *The Scientist's Guide to Writing*. Princeton, NJ: Princeton University Press. https://doi.org/10.2307/j.ctvcmxs67

Herbert, Frank. 1965. *Dune*. New York: Ace Books. https://archive.org/details/dune0000herb_a7n1/mode/2up

Hey, Tony, Stewart Tansley, and Kristin Tolle, eds. 2009. *The Fourth Paradigm: Data-Intensive Scientific Discovery*. Seattle: Microsoft Research. https://www.microsoft.com/en-us/research/publication/fourth-paradigm-data-intensive-scientific-discovery/

Huang, Dongmei, Wei Song, and Guoliang Zou. 2019. *Marine Big Data*. Hackensack: World Scientific. https://doi.org/10.1142/11337

Huang, Dongmei, Danfeng Zhao, Lifei Wei, Zhenhua Wang, and Yanling Du. 2015. "Modeling and Analysis in Marine Big Data: Advances and Challenges." *Mathematical Problems in Engineering*. https://doi.org/10.1155/2015/384742

Milroy, Scott. 2016. *Field Methods in Marine Science: From Measurements to Models*. New York: Garland Science. https://www.routledge.com/Field-Methods-in-Marine-Science-From-Measurements-to-Models/Milroy/p/book/9780815344766#

Montgomery, Thomas D., Joanne Rae Buchbinder, Ellen S. Gawalt, Robbie J. Iuliucci, Andrew S. Kock, Evangelia Kotsikorou, Patrick E. Lackey, Min Soo Lim, Jeffrey Joseph Rohde, Alexander J. Rupprecht, Matthew N. Srnec, Brandon Vernier, and Jeffrey D. Evanseck. 2022. "The Scientific Method as a Scaffold to Enhance Communication Skills in Chemistry." *Journal of Chemical Education*. https://doi.org/10.1021/acs.jchemed.2c00113

Murray, John. 1895. *Report on the Scientific Results of the Voyage of the H.M.S. Challenger During the Years 1872–76: A Summary of the Scientific Results, First Part*. Published by Order of Her Majesty's Government. https://doi.org/10.5962/bhl.title.6513

National Academies of Sciences, Engineering, and Medicine. 2017. *Communicating Science Effectively: A Research Agenda*. Washington, DC: National Academies Press. https://doi.org/10.17226/23674

National Research Council. 2002. *Scientific Research in Education*. Washington, DC: National Academies Press. https://doi.org/10.17226/10236

Newton, Sir Isaac. 1675. "Letter to Robert Hooke." https://digitallibrary.hsp.org/index.php/Detail/objects/9792

Popper, Karl. *The Logic of Scientific Discovery*. 2nd Ed. Milton Park: Routledge. https://www.routledge.com/The-Logic-of-Scientific-Discovery/Popper/p/book/9780415278447

Rentzsch, Katrin, and Michaela Schröder-Abé. 2014. "Self-Enhancement 2.0: An Integrated Approach to Measuring Dyadic Self-Enhancement at Two Levels." *Social Psychological and Personality Science* 6(3): 251-258. https://doi.org/10.1177%2F1948550614558634

Rudzin, Johna E., Dax C. Soule, Justine Whitaker, Halle Berger, Sophie Clayton, and Kristen E. Fogaren. 2022. "Catalyzing Remote Collaboration During the COVID-19 Pandemic and Beyond: Early Career Oceanographers Adopt Hybrid Open Science Framework." *Frontiers of Marine Science* 9: 1–7. https://doi.org/10.3389/fmars.2022.855192

Warren, Bruce A. 2006. "Historical Introduction: Oceanography of the General Circulation to the Middle of the Twentieth Century." In *Physical Oceanography: Developments Since 1950*. Edited by Jochum Markus and Raghu Murtugudde, 1–14. New York: Springer. https://doi.org/10.1007/0-387-33152-2

West, Darrell M., and John Allen. 2018. "How Artificial Intelligence is Transforming the World." Brookings. April 4, 2018. https://www.brookings.edu/research/how-artificial-intelligence-is-transforming-the-world/

Figure 2.26
Fullerton College students pose for an obligatory sea cave photo after exploring the tidepools at Crystal Cove State Beach in Laguna Beach, California. Field trips like these inspire curiosity and perhaps even an interest in pursuing science as a career.

Credit: Chamberlin

You can sleep when you're dead.
Artwork by Martha Rich.

Science Aboard Ships

3

Figure 3.1

Sunken ships attract marine life and create reefs that support abundant and diverse life forms.

Credit: istock/ Nicolas Sanchez-Biezma

We can never know for sure who was the first person to dive beneath the waves in search of food or artifacts, but we do know that ancient peoples—perhaps as far back as one hundred thousand years ago—collected shells, shellfish, and other valuable undersea resources along the coastline (e.g., Will et al. 2019). As human use of the ocean grew for military, economic, and industrial purposes, so too did the need to know more about the ocean. Studies of marine biology, oceanography, and fisheries emerged as scientific quests to acquire knowledge of the sea for human purposes (e.g., Mills 2012).

In the next three chapters, we explore the approaches and technologies used to carry out ocean science—the methods of an oceanographer. Admittedly, a tour of methods might seem a bit like wandering through a home improvement store. (Indeed, oceanographers often shop at these stores—you can never have enough duct tape on board a rolling ship.) But a brief survey of these technologies will help you appreciate the myriad ways in which oceanographers are probing the world ocean. This appreciation will help you better understand the team approach to ocean science and the interconnectedness of processes in the world ocean.

Figure 3.2
The Chumash people have used tomols—wood plank canoes waterproofed with tar—to travel the waters of Southern California and the Channel Islands for more than 10,000 years.

Credit: Robert Schwemmer, Channel Islands National Marine Sanctuary, National Ocean Service, NOAA CC BY 2.0

My friend and colleague University of California (UC) Santa Barbara physical oceanographer Tommy Dickey (*b*. 1945) spent his career pushing the boundaries of ocean science through application of new technology. In his 2020 paper "New Discoveries Enabled by the Emergence of High-Resolution, Long-Term Interdisciplinary Ocean Observations," he wrote:

> The vastness and inhospitable nature of our oceans has limited our ability to understand oceanic phenomena and to make ocean predictions. Great progress in sampling the vertical and temporal variability of the ocean was made with new sensors and moorings in the period of 1980–2000; however there remain major challenges to adequately sample the full three-dimensional and temporal variability of the interior ocean.

3.1 The Birth of Oceanography

While many important scientific expeditions took place in the 19th century, one above all, the *Challenger* expedition (1872–1876), brought together the interdisciplinary approaches that would characterize the field of oceanography in modern times. Oceanographer and marine biologist Henry Bigelow (1879–1967), who helped found the world-famous Woods Hole Oceanographic Institution (WHOI) in Massachusetts, wrote in his 1931 textbook, *Oceanography*, "Students of the history of science may well date the birth of modern oceanography from December 21, 1872, the date when [HMS *Challenger*] put to sea from Plymouth, England." Indeed, the *Challenger* expedition was the first devoted entirely to ocean science. A decade after their voyage, Scottish expedition leader Sir Charles Wyville Thomson (1830–1882) and Canadian oceanographer John Murray (1841–1914) wrote:

> The vast ocean lay scientifically unexplored. All the efforts of the previous decade had been directed to the strips of water round the coast and to enclosed or partially enclosed seas. . . . This consideration led to the conception of the idea of a great exploring expedition which should circumnavigate the globe, find out the most profound abysses of the ocean, and extract from them some sign of what went on at the greatest depths. (Thomson and Murray 1885, p. 1)

Figure 3.3
The *Challenger* expedition established many of the routines used by modern oceanographers.

Credit: Shaw and Chamberlin

The *Challenger* expedition covered 79,277 miles (127,584 km) of ocean and gained fame for its ambitious plan and the magnitude of its accomplishments. Its results filled 50 printed volumes. But perhaps most lasting among its contributions was its influence on the way oceanographers sample the ocean, a sampling approach that remains an integral part of oceanographic research more than 150 years later.

3.2 Sampling over Space

Since the days of the *Challenger* expedition, shipboard oceanography has used a sampling approach based on grids of points stretched over a region of interest. As Thomson describes it, "At intervals as nearly uniform as possible, we established 362 observing stations" (Thomson 1878, p. xi). Thomson's observing stations are now known as **hydrographic stations**, or simply stations. Each point on the grid represents a station. The line along which the ship moves from station to station is known as a **transect**. Hydrographic stations mark the spots where geological, chemical, physical, and biological measurements and samples may be taken. Of course, oceanographers may carry out measurements and take

Figure 3.4
By sampling at regular intervals along a grid of stations, oceanographers obtain a snapshot of the ocean's properties across a geographic region.

Credit: Shaw and Chamberlin

samples along the transect while underway, too. The number of stations and length of the transect depend on the number of days allotted to the expedition and the type and amount of work that the expedition team desires to carry out.

The distance between samples of different regions of the ocean—horizontally or vertically—defines the **spatial resolution** of the measurements. As someone living in the 21st century, you have undoubtedly witnessed increases in the spatial resolution of televisions and smartphone cameras. Spatial resolution defines the smallest object or feature that may be resolved or

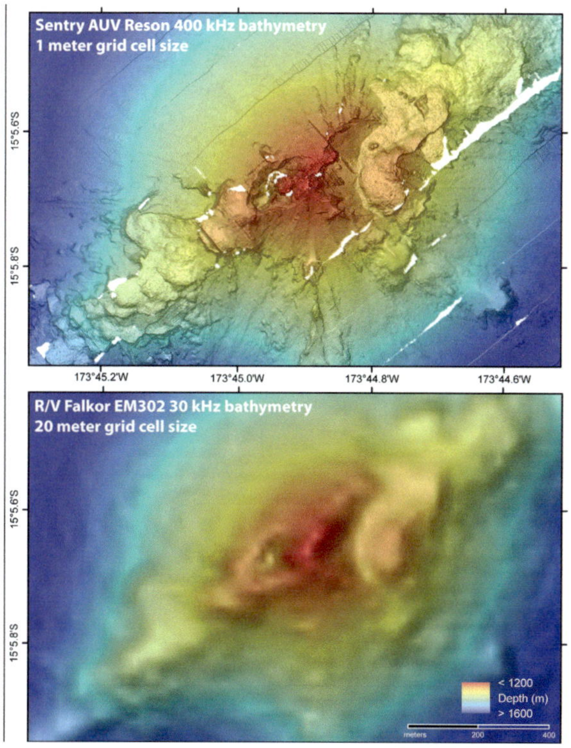

Figure 3.5
Spatial resolution controls the amount of detail available in an image or a set of oceanographic measurements. Note the improved detail in the 1 meter grid image (top) versus the 20 meter grid (bottom).

Credit: Courtesy of Schmidt Ocean Institute

distinguished by the display or sensor in your television or smartphone camera—such as 4K resolution (and soon 8K and 16K). Similarly, for oceanographers, spatial resolution refers to the smallest features that may be distinguished in a particular set of data. At higher spatial resolution, oceanographers can identify an individual underwater mountain versus a blob of mountains. They can discriminate the fine details of different water temperatures (depicted as colors) instead of one blotch of the same temperature. Low spatial resolution occurs when oceanographers have to take samples far apart in distance or when a sensor can only measure broad swaths of the ocean. To obtain higher spatial resolution, oceanographers must take more samples in a given area or build sensors that detect smaller features. But there are limits.

Time, technology, and funding don't always permit the highest spatial resolution. A compromise has to be made between *where* oceanographers sample and *how much* they sample: more locations with fewer samples at each location or fewer locations and more samples. Taking and processing more samples takes more time. Observing smaller details of the ocean demands better technology. But time and technology require money that is not always available.

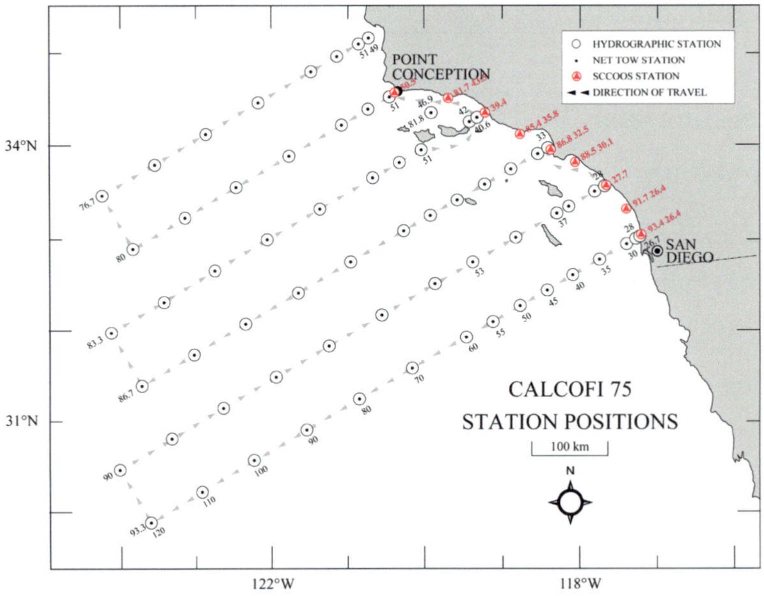

Figure 3.6

The station map for the California Cooperative Oceanic Fisheries Investigation includes 75 hydrographic stations.

Credit: Scripps Institution of Oceanography (SIO)-CalCOFI

Figure 3.7 CalCOFI oceanographers prepare to deploy a plankton net off the California coast.

Credit:* © *James R Wilkinson/SIO-CalCOFI

An excellent example of the application of grid sampling can be found in the California Cooperative Oceanic Fisheries Investigation (CalCOFI), which has been conducting oceanographic surveys off the coast of California since 1949. The original grid in 1950 included hundreds of stations along the California coastline and parts of Baja. But since 2004 the survey has been limited to between 75 and 109 stations over a smaller area. Today CalCOFI samples an area of 73,184 square miles (189,547 km^2), a considerable expanse but still less than 0.05 percent of the world ocean. Nevertheless, CalCOFI ranks as "the longest-running oceanographic monitoring program on the planet" (e.g., Koch 2019).

Combinations of technology and ships now provide an ability to sample the ocean at spatial scales spanning twelve orders of magnitude—from kilometers down to nanometers (e.g., Dickey 1990; Branson et al. 2016). The challenge for 21st-century oceanographers will be to connect the dots—to find the connections—between processes at each of those scales (e.g., Dickey 1991).

Oceanographers desire high-resolution sampling in space and time. But sampling large regions quickly and repeatedly proves challenging within the constraints of ship time, technology, and funding.

3.3 Sampling over Time

Ocean properties also vary with time. Similar to spatial resolution, we refer to **temporal resolution** (i.e., time resolution) as the interval of time between samples. A once-a-month sampling program has a temporal resolution of one month. A daily sampling regime has a temporal resolution of a day, and so on. But why is this important?

Imagine that you were tasked with summarizing the details of every movie playing in a 21-screen multiplex, all with different start times. To complete the task, you would have to move from theater to theater and watch only a few minutes of one movie before moving on to the next. You would never be able to see one movie in its entirety. And if you spent more time on one movie, you would sacrifice your knowledge of other movies.

The ocean is like a multiplex in which thousands—if not millions—of movies are playing simultaneously. What's more, the plot of each movie keeps changing as different characters and events arise. When you return to the first theater, there's no guarantee that the original film will still be playing. It's maddening, but somehow, through persistence, repetition, well-honed methods, and careful sampling strategies, oceanographers have begun to make sense of it all.

Figure 3.8 Snapshots of the same place over time may reveal dramatic differences. A sample taken only at 11 a.m. would have missed the octopus entirely.

Credit: Shaw and Chamberlin

In the CalCOFI program, oceanographers obtain one set of samples per station approximately every three months. Such sampling permits oceanographers to track changes on seasonal scales, but what about day-to-day or week-to-week changes? Oceanographers simply don't have the tools or manpower to sample the ocean over all relevant scales of space or time. But, as we shall see in the chapters ahead, new technologies are helping to improve ocean sampling at ever-finer spatial and temporal scales.

3.4 The Challenge of Pressure

Nearly every technology developed for ocean exploration beneath its surface confronts one inevitable question: How deep can it go without being crushed? Many a machine has been lost (some with lives aboard) because of the intense pressure exerted by the ocean. Pressure is the reason why it's easier to explore outer space. A dozen humans went to the Moon half a century ago, traveling a distance of 238,855 miles (384,400 km), but until 2019, only three humans had been to the **Challenger Deep**, the deepest location in the world ocean, a mere 6.8 miles (10,925 m) deep. Though dozens of people have now descended to the Challenger Deep (see below), extreme water pressure remains a barrier to all but the most specialized technologies.

Physicists define **pressure** as the force exerted by a fluid (a gas or a liquid) on an object immersed within it. In the ocean, every 10 meters of water depth, about 33 feet, increases pressure by about 14.7 pounds per square inch, or 1 atmosphere (atm) at sea level. Earth's atmosphere, on average, extends about

Figure 3.9 Styrofoam cups, placed aboard an instrument lowered deep in the ocean, illustrate the effects of pressure.

Credit: NOAA CC BY 2.0

7.5 miles (9.17 km) above us. So the pressure exerted by about 33 feet of water equals the pressure exerted by about 7.5 miles of gases in our (one) atmosphere. For simplicity, oceanographers often express water pressure in units of atmospheres. The calculation is quite simple:

$$P\ (\text{atm}) = z\ (\text{m}) \ /\ 10\ (\text{m/atm}) + 1\ (\text{atm})$$

(***Eq. 3.1***)

where P is water pressure, z is depth, and 1 atm represents the pressure at sea level (from the atmosphere pressing down on the ocean).

The simple math makes it apparent: you don't have to go very deep to experience great pressure. At 800 times the density of air, water exerts a much greater pressure on objects immersed within it. At a depth of a hundred feet (roughly 30 m), a diver experiences three times the pressure they would experience at sea level. Free divers—persons who dive with no scuba gear and rely only on holding their breath—experience even greater pressure. The current no limits freediving record depth of 702 feet (214 m), held since 2007 by Austrian Herbert Nitsch (*b*. 1970), converts to a water pressure 22 times the pressure at sea level (e.g., The Salt Sirens 2021). Imagine a newborn baby elephant (about 300 lbs) balanced on a USB wall charger (about 1 square inch on each side). That's what a free diver experiences at those depths.

Figure 3.10
At 700 feet, water pressure is equivalent to the weight of a baby elephant on every inch of a diver's body.

Credit: Shaw and Chamberlin

3.5 The Undersampling Problem

The challenges of sampling a difficult environment across multiple scales of space and time have led to what is known as the **undersampling problem** in oceanography. Oceanographers do not yet possess a sufficient number of measurements to develop a complete understanding of ocean processes. Advances in technology—especially satellites and robots—have improved the spatial and temporal resolution of observations. Though some oceanographers contend the physics of the ocean is adequately sampled (e.g., Munk 2000), many parts of the ocean, especially the deep ocean, and many biological processes remain undersampled (e.g., van Haren 2018; Claustre et al. 2020).

Figure 3.11 Many parts of the world ocean remain unexplored and undersampled.

Credit: **NOAA CC BY 2.0**

3.6 Modern Oceanographic Vessels

Oceanographic research vessels form the backbone of ocean research in the 21st century. While their ability to provide much-desired spatial and temporal resolution remains limited, ships offer a platform for nearly every other type of technology used to study the ocean. Humans working aboard ships can perform feats no modern robots can come close to. Ships enable diverse teams of oceanographers from all over the world to carry out collaborative research.

Modern oceanographic research vessels—designated by R/V in front of the ship's name—serve several functions. Under the supervision of a trained captain and crew, they provide a safe, comfortable, mobile habitat for dozens of scientists and technicians. Among their many attributes,

Figure 3.12
The R/V *Sally Ride*, a multidisciplinary oceanographic research vessel commissioned in 2016, was named after the first American woman in space.

Credit: *Office of Naval Research* CC BY 2.0

1. They provide a stable platform for launching and deploying a variety of oceanographic instruments above, below, and on the surface of the ocean.
2. They permit scientists to access remote ocean locations, including the central gyres of the major oceans, ice-covered oceans, and remote islands.
3. They offer state-of-the-art navigation capabilities, providing precise locations for sampling (underway or on station) and the ability to return to those exact locations in the future.
4. They offer state-of-the-art communications and telepresence capabilities, allowing oceanographers and crew to download satellite imagery and stream images to shoreside personnel and the public.
5. They provide specialized cranes and winches for deploying oceanographic instruments, such as submersibles and robotic craft.
6. They offer ample deck space, such as a **fantail**, a broad platform on the rear of a vessel, where oceanographers can work outside close to the waterline, and where portable vans (or clean rooms) can be installed for carrying out specialized analyses and experiments.
7. They provide laboratories with benchtop instruments for at-sea analyses of collected samples, where live organisms may be observed, and where samples may be prepared and stored for shoreside analyses.

Figure 3.13 Oceanographers direct nighttime recovery of a side-scan sonar aboard the R/V *Point Sur* in the Gulf of Mexico in August 2021.

***Credit:* NOAA Ocean Exploration Cooperative Institute, Inner Space Center, and Patrick Flanagan**

8. They offer computing capability for storing and analyzing electronic data in real time or on demand.
9. They provide mess halls and places to sleep, bathe, and relax (e.g., a lounge with a big-screen TV, a library, a gym, and, on some vessels, a sauna). Some of these spaces can also be used for meetings and conversations with colleagues.

Most oceanographic research vessels operate under the supervision of the University–National Oceanographic Laboratory System, better known as UNOLS. The UNOLS fleet includes five global-class ships, six intermediate-range oceangoing vessels, and seven regional/coastal-class ships. UNOLS also coordinates the use of nearly one hundred small research vessels (35–100 feet in length). UNOLS-approved vessels meet strict standards, including rules aimed at individual conduct, prohibitions on drugs and alcohol, and guidelines on privacy, discrimination, and harassment while on board (UNOLS 2021). Researchers who request use of a UNOLS vessel must adhere to these standards.

Oceanographers also make use of **ships of opportunity**, commercial vessels that regularly cross the ocean, such as freighters, tankers, cruise ships, and ferries. The Ships of Opportunity Program (SOOP) provides

Figure 3.14

It's not Caesar's Palace, but the R/V *Polarstern* is pretty luxurious for a research vessel.

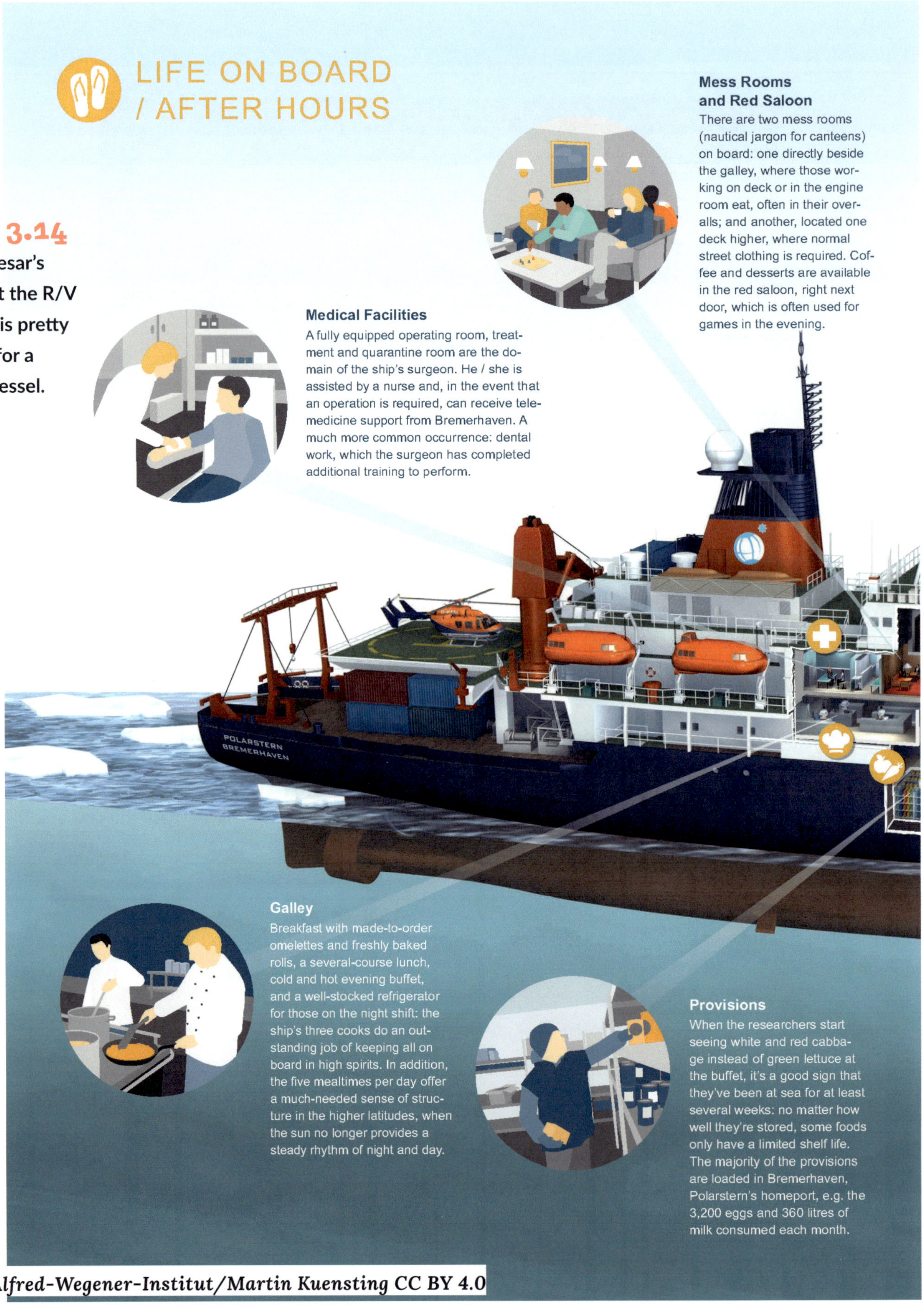

MOSAiC
International Arctic Drift Expedition

Produced by AWI GRAPHIC © Martin Künsting

Library / Blue Saloon

Polarstern officially belongs to the Federal Republic of Germany, represented by the Federal Ministry of Research. Receptions for honoured guests from abroad in international ports, as well as festive addresses during expeditions, take place in the blue saloon. It also features a branch of the Alfred Wegener Institute's library, which includes a wealth of scientific literature, as well as historical works on polar research.

Staterooms

In terms of accommodations, every researcher on board shares a roughly ten-square-metre stateroom (nautical jargon for room) and bathroom with a roommate. For a bit of privacy, and to get some sleep, they can draw a curtain in front of their bunk. Discipline is called for in terms of how much each participant brings with him / her, and in day-to-day life, since storage space is limited.

Computer Room

Processing data, printing, or sorting images: the computer room is the place to go when researchers don't want to disturb their roommates with their work. The room is also home to one computer connected to the Internet for the ca. 100 participants – the limited ability to transfer data via satellite in the polar regions makes this necessary. But researchers can also send emails from their own devices using an internal network.

Recreation Area

After a day working under Arctic conditions, many participants enjoy thawing out in the sauna. Or they can use the ten-by-five-metre swimming pool, where water polo is the game of choice. Those who've spent most of their day sitting at a computer can get some exercise on the rowing machine, while a solarium helps to make up for the lack of light during the long Polar Night.

reports on weather conditions and deploys expendable profilers that measure properties of the **water column**, an unspecified volume of water from the surface to a particular depth. SOOP, managed by the National Oceanic and Atmospheric Administration (NOAA), currently collects weather and ocean data from more than 1,200 vessels.

It goes without saying that life aboard an oceanographic research vessel brings adventure, excitement, and a sense of camaraderie with fellow scientists. Crew and scientists typically work around the clock in shifts of four hours on and eight hours off or some variation thereof. With multiple teams of scientists carrying out different kinds of investigations, a research cruise may seem a little like a three-ring circus. But in the end, weather and equipment permitting, everyone accomplishes as much as they possibly can in the time allotted. It's a lot of work, and the work can be exhausting, but you won't get any sympathy from the crew. When I once complained on the R/V *Oceanus* about being tired, the galley person responded, "You can sleep when you're dead."

Figure 3.15
On Thursdays aboard the *Polarstern*, you get cake (if you've been a good oceanographer). Mealtimes offer a chance to interact with the crew and other scientists, or simply take a moment's break from the grueling shipboard routine.

Credit: Alfred-Wegener-Institut/Esther Horvath CC BY 4.0

3.7 Beneath the Surface

Technology that permits humans to travel beneath the surface of the ocean dates back at least to the fourth century BCE. The Greeks apparently figured out they could hoist a large metal pot over their head, walk out into the water, and breathe the air trapped inside. Thus was born the **diving bell**, a dome-shaped, air-filled chamber that permits divers to work at depth, even in modern times. Aristotle (384–322 BCE) mentions diving bells used by sponge divers in the Aegean Sea (Forster 1927). His student, Alexander the Great (356–323 BCE), purportedly used a diving bell to observe a "gigantic creature that took three days to pass by" (Bevan 1999). Of course, had he stayed submerged for such a time, he would have been dead from lack of oxygen. Obviously, you can't believe everything you read in books.

Human desire to reach the deepest depths of the ocean finds a place in the 20th century as well. In the 1930s, two Americans, naturalist William Beebe (1877–1962) and engineer Otis Barton (1899–1992), built the world's first **bathysphere**, a hollow steel sphere with transparent baseball-sized portholes for making observations. Lowered on a cable from a ship, the bathysphere—with the men bolted inside—made a series of successful descents, including one to 3,028 feet (923 m) on August 11, 1934. Beebe's sketches and account of his dives resulted in a best-selling book, *Half Mile Down* (1934), the first work to document the strange and wonderful animals that inhabit the dimly

Credit: NOAA/University of Alaska Fairbanks/ Global Explorer Oceaneering CC BY 2.0

Figure 3.16 One of many awe-inspiring animals in the mesopelagic zone, often called the twilight zone.

Figure 3.17
You never know who might come knocking when you visit the deep sea.

Credit: Shaw and Chamberlin

lit waters known as the **mesopelagic zone**.

A few decades later, a similar vehicle, the **bathyscaphe** *Trieste*, enabled two men to reach the Challenger Deep, the first humans ever to do so. A bathyscaphe consists of a hollow steel sphere attached to a giant gasoline-filled "bladder," which provides the necessary buoyancy for returning divers to the surface without a cable. On January 23, 1960, Swiss oceanographer and engineer Jacques Picard (1922–2008) and US Navy lieutenant-oceanographer Don Walsh (*b*. 1931) descended to 35,814 feet (10,916 m), setting a world record for the deepest dive. Their record would stand for 52 years (Walsh 1962).

3.7.1 SCUBA

No summary of underwater technology would be complete without mention of the invention that permitted humans to see the undersea world firsthand: the self-contained underwater breathing apparatus, better known as **scuba**. Invented in 1943 by French engineer Émile Gagnan (1900–1979) and French ocean explorer Jacques Cousteau (1910–1997), scuba permits humans to inhabit shallow waters for limited periods of time. Cousteau's television series, *The Undersea World of Jacques Cousteau* (1968–1975), instilled in the public a love for the ocean

Figure 3.18
Scuba is a popular and safe sport for anyone who wishes to visit the fishes.

Credit: istock/Mihtiander

and inspired the careers of many a marine biologist and oceanographer, including mine (Cousteau and Schiefelbein 2007).

While typically limited to shallow waters (less than 130 ft), **scientific diving**—underwater observations, measurements, and experiments using scuba—provides a means for scientists to carry out direct observations or experimental studies for extended periods of time. In recent years, marine scientists have begun to use advanced scuba tools (i.e., advanced technical diving) to explore **mesophotic coral ecosystems**, deep, light-dependent coral reefs found at depths between 100 and 500 feet (30–250 m; e.g., Loya et al. 2019; Bell et al. 2022).

Scientific divers may also employ a **shark cage** (also invented by Cousteau)—a metal cage that permits safe viewing of sharks and other pelagic ocean species. Divers enter and exit the cage at the surface through an opening in the top. Once divers are inside, the cage descends to an optimal depth for viewing the animals. Though shark-cage diving is popular as a tourist adventure, the cages prove useful for scientific studies where direct observation is required, including estimates of

Figure 3.19
Shark cages provide an ideal platform for observing sharks' natural behavior.

Credit: istock/ShayneMyersPhoto

animal length (e.g., May et al. 2019) and studies of feeding behavior (e.g., Becerril-Garcia et al. 2019). I'm told there's nothing like meeting a great white shark face-to-face. Maybe some day you'll get the opportunity.

3.7.2 Atmospheric Diving Suits

The **atmospheric diving suit**, or ADS—also known as an armored diving suit or articulated diving suit—allows divers to explore deeper waters. The ADS resembles a space suit, but it operates more like a submersible (see below). In fact, some authors classify the ADS as a one-person submersible. The ADS maintains sea-level pressure inside the suit—hence its name. This eliminates the need for decompression when the diver comes to the surface.

An early type of ADS, the JIM suit, enabled American marine biologist Sylvia Earle (*b*. 1935) to reach 1,281 feet (381 m) off the coast of Oahu in 1979. The dive lasted two and a half hours and set a record for the deepest dive at the time. During her dive, Earle observed bioluminescence in bamboo corals, a phenomenon not filmed until 2016 (Helgason 2016). The JIM suit—named after a diver named Jim—was also featured in the

James Bond film *For Your Eyes Only* (Glen 1981) and a lesser-known sci-fi film, *DeepStar Six* (Cunningham 1989).

New-generation ADSs can reach depths of 2,000 feet (610 m). They incorporate various types of propulsion units and include LED lights, sonar, HD cameras, real-time atmospheric monitoring, and two-way wired and acoustic communication systems. In 2019, the US Navy teamed up with the National Aeronautics and Space Administration (NASA) to develop a diver-augmented vision device, a heads-up display system for dive helmets. Using the display, the diver can monitor a host of data and receive text messages from topside (such as ROSL, roll on seafloor laughing). The system can even provide augmented-reality displays, useful where visibility is poor (Melnick 2019; Naval Technology 2021).

Credit: *Shaw and Chamberlin*

Figure 3.20 **Though largely used for commercial and military applications, the ADS holds great potential for ocean science.**

3.7.3 Submarines

Though predominantly used for military purposes, **submarines**, defined as a vessel capable of propelling itself underwater, also play a role in modern oceanographic research. Since 1993 oceanographers and the US Navy have collaborated on the use of nuclear submarines for oceanographic research, especially in the Arctic. The first mission, named SCientific ICe EXercise 1993, or SCICEX-93, carried five scientists aboard the USS *Pargo*, a Sturgeon-class fast attack submarine

Figure 3.21
The USS *Connecticut* breaks ice in the Arctic. Submarines play a minor yet important role in ocean science, going where few other craft—manned or unmanned—can go.

Credit: **US Navy/Daniel Hinton CC BY 2.0**

named after a fish. The success of SCICEX-93 spurred five more missions from 1995 to 1999. For security reasons the navy instituted scientific accommodation missions, or SCICEX SAMs, in 2000, in which trained navy personnel carried out oceanographic measurements without scientists on board. Scientists may access the data—once it's declassified—through the National Snow and Ice Data Center (SCICEX Science Advisory Committee 2010). As a joint civilian-military program, SCICEX SAMs enable collection of data that would not otherwise be possible to obtain. Observations beneath the ice help verify above-ice satellite observations and further scientific understanding of the role of the Arctic in climate change (Morelo 2010).

The US Navy has long supported oceanographic research, primarily for its applications to surveillance, defense, and warfare. See Oreskes 2020.

3.7.4 Submersibles

Small submarines, better known as **submersibles**, deep submergence vehicles (DSVs), or human-occupied vehicles (HOVs), play a larger role in oceanographic research than submarines, sacrificing speed for depth capability. Unlike submarines, submersibles require a transport ship to carry them to their dive location. And while typical naval submarines carry more than a hundred personnel, most submersibles carry only two or three. Typical modern submarines operate at depths less than 1,500 feet (457 m), but many submersibles can attain depths of 14,763 feet (4,500 m) or more. New generations of submersibles can now reach the deepest locations in the world ocean.

No submersible in the world can match the record of WHOI's HOV *Alvin*—in operation since 1964—the only American scientific research submersible. *Alvin* has enabled the discovery of hydrothermal vents, helped find the RMS *Titanic*, and served countless scientists in pursuit of knowledge about the ocean (e.g., Humphris et al. 2014). As of 2020, the near-60-year-old vessel had logged 5,065 dives, carried 15,186 people, and spent 35,474 hours submerged. But *Alvin*'s most exciting days may lie ahead. In 2022, upgrades to the submersible allowed it to reach a

Figure 3.22
HOV *Alvin* prepares to dive from the fantail of the R/V *Atlantis*.

Credit: NOAA/Bureau of Ocean Energy Management/USGS CC BY 2.0

Figure 3.23
On July 21, 2022, *Alvin* reached its deepest dive ever. Not bad for a 60-year-old sub.

Credit: *Shaw and Chamberlin*

depth of 21,325 feet (6,453 m), the deepest ever for the sub. It may now access some 98 percent of the seafloor. Despite the opinion of some that robots could better serve science at these depths, there's simply no substitute for being there. As one ocean engineer put it, underwater cameras are "still a long way short of what the human eye can do" (Oberhaus 2020).

Still, the very deepest spots remain mostly unreachable. Only two submersibles, the Chinese *Fendouzhe* and the US DSV *Limiting Factor*, can reach the Challenger Deep today. Descending in the *Limiting Factor* in April 2019, American undersea explorer Victor Vescovo became the fourth explorer to reach the Challenger Deep and set a new world record for

Figure 3.24
R/V *Atlantis* swimmer talks to the *Alvin* crew following a successful dive.

Credit: NOAA/*Bureau of Ocean Energy Management*/USGS CC BY 2.0

the deepest dive ever, reaching 35,853 feet (10,928 m). On July 12, 2022, American marine geologist Dawn Wright became the first Black person (and fifth woman) to reach the deepest spot on Earth with a 35,823-foot dive (10,919 m) aboard the *Limiting Factor* (Weinman 2022). More than two dozen people have now reached these depths (Kreier 2022).

3.7.5 Undersea Research Stations

Undersea research stations—semipermanent structures designed for human habitation and scientific research on the seafloor—entered the human imagination long before they became a reality. Historians credit "freethinking" English bishop John Wilkins (1614–1672) as the first person to imagine permanent undersea colonies (Wilkins 1708; Matsen 2009). French writer Jules Verne (1828–1905) imagined a mobile underwater habitat in his epic novel *Twenty Thousand Leagues Under the Sea* (Verne 1870; translated by Miller and Walter 1993). Captain Nemo's exploits aboard *Nautilus* inspired many an ocean explorer, including Cousteau (Matsen 2009).

In the 1960s Cousteau and his team of divers built three stationary underwater living spaces on the seafloor of the Mediterranean and the

Figure 3.25 Humans have long dreamed of living underwater. Perhaps one day that dream will become a reality.

Credit: istock/ratpack223

Figure 3.26
The Conshelf II submarine garage—as it appeared in 2012—now offers a habitat for fishes and other marine life.

Credit: Flickr/Wreckferret CC BY-SA 2.0

Red Sea. The first, the cylindrical Continental Shelf Station Number One, or simply Conshelf I, was established at a depth of 33 feet (10 m) near Marseille, France. At 17 feet long and 8 feet high (about 5.2 × 2.4 m), it served as a kind of underwater tiny house complete with television, radio, and a record player. From September 14 through 21, 1962, French aquanauts Albert Falco (1927–2012) and Claude Wesley (1930–2016) spent a week on Conshelf I, the first humans ever to live underwater for multiple days. According to protocol, the men were required to spend five hours a day outside of the station carrying out observations of marine life and tending an underwater farm. When asked what he would like to do upon his return to the surface a week later, Falco simply replied, "To walk" (Cousteau and Dugan 1963, p. 324).

A year later, the Conshelf II "village" was established in the Red Sea. The starfish-shaped main station, at 33 feet long (10 m), accommodated five men and included a parrot to warn of pressure changes. The undersea village also had a garage for a submersible, an equipment hangar, and a second station 82 feet deeper (25 m). This time, the

Figure 3.27 Aquarius Reef Base aquanauts pose for pictures during the NASA Extreme Environment Mission Operations (NEEMO) 21 in 2016.

Credit: NASA/Mark Widick

aquanauts stayed a month. During the last four days of their residency, they were joined by Cousteau's research partner and wife, Simone Cousteau (1919–1990), who became the first woman to live underwater.

The exploration and research in these underwater habitats were captured in the documentary *World without Sun* (Cousteau 1964), which won an Academy Award in 1965 (Cousteau's second in the category). That year the Cousteau team also established a much deeper habitat near Nice, France. Conshelf III, at 336 feet (102 m), housed six aquanauts who remained underwater for three weeks. The idea of people living and working beneath the sea was no longer a fantasy. Cousteau had made it a reality.

Cousteau's success led to construction of similar habitats around the world. Since 1962 at least 70 underwater habitats have been built (Chamberland 2007). But don't let that number fool you—many never made it beyond their first mission. Most no longer exist or have become manmade reefs. And only three ever accommodated aquanauts for longer than 30 days.

The only remaining undersea habitat dedicated to scientific research is Florida International University's Aquarius Reef Base (1986–present), a 400-square-foot, six-person habitat in the Florida Keys. One other operational undersea habitat, the Jules Undersea Lodge (formerly La Chalupa research laboratory), serves as an underwater hotel and occasional classroom in a lagoon in Key Largo Undersea Park in Florida. In 2023, University of South Florida professor Joseph Dituri (*b*. 1968) set a world record for living underwater in the Jules Undersea Lodge (Kim 2023).

In 2020, Cousteau's grandson, French ocean explorer Fabien Cousteau (*b*. 1967), and Swiss designer Yves Béhar (*b*. 1967) announced plans to build an underwater manned research station in 60 feet of water (about 18 m) off the coast of Curaço in the Caribbean. Named PROTEUS™, the planned habitat will accommodate a dozen aquanauts and have a capability to live-stream video in 16K resolution. About the project Cousteau says, "I'm just a crazy person with a dream that sees this as being not only possible—but absolutely necessary—for our future well-being" (Knapp and Jennings 2020). A partnership announced with NOAA in May 2023 may just help that dream come true (NOAA 2023).

Figure 3.28
The PROTEUS™ habitat promises to launch a new era in human habitation of the sea.

***Credit:* PROTEUS™**

3.8 Chapter References

Becerril-Garciá, Edgar E., Daniela Bernot-Simon, Marcial Arellano-Martínez, Felipe Galván-Magaña, Omar Santana-Morales, and Edgar M. Hoyos-Padilla. 2020. "Evidence of Interactions between White Sharks and Large Squids in Guadalupe Island, Mexico." *Scientific Reports* 10: 17158. https://doi.org/10.1038/s41598-020-74294-4

Beebe, William. 1934. *Half Mile Down*. New York: Harcourt, Brace, and Company. https://doi.org/10.5962/bhl.title.10166

Bell, James J., Valerio Micaroni, Benjamin Harris, Francesca Strano, Manon Broadribb, and Alice Rogers. 2022. "Global Status, Impacts, and Management of Rocky Temperate Mesophotic Ecosystems." *Conservation Biology*, e195235. https://doi.org/10.1111/cobi.13945

Bevan, John. 1999. "Diving Bells Through the Centuries." *South Pacific Underwater Medicine Society Journal* 29(1): 42–50.

Bigelow, Henry B. 1931. *Oceanography: Its Scope, Problems, and Economic Importance*. Boston: Houghton Mifflin. https://hdl.handle.net/1912/219

Branson, Oscar, Elisa A. Bonnin, Daniel E. Perea, Howard J. Spero, Zihua Zhu, Maria Winters, Bärbel Hönisch, Ann D. Russell, Jennifer S. Fehrenbacher, and Alexander C. Gagnon. 2016. "Nanometer-Scale Chemistry of a Calcite Biomineralization Template: Implications for Skeletal Composition and Nucleation." *Proceedings of the National Academy of Sciences* 113(46): 12934-12939. https://doi.org/10.1073/pnas.1522864113

Chamberland, Dennis. 2007. *Undersea Colonies: The Future of Permanent Undersea Settlements*. Orlando: Quantum Editions.

Claustre, Hervé, Kenneth S. Johnson, and Yuichiro Takeshita. 2020. "Observing the Global Ocean with Biogeochemical-Argo." *Annual Review of Marine Science* 12: 23–48. https://doi.org/10.1146/annurev-marine-010419-010956

Cousteau, Jacques-Yves. 1964. *World without Sun*. Columbia Pictures. 93 minutes. https://www.imdb.com/title/tt0058364/

Cousteau, Jacques-Yves, and James Dugan. 1963. *The Living Sea*. New York: Harper & Row. https://www.google.com/books/edition/the_living_sea/RH62lplz-2AC

Cousteau, Jacques-Yves, and Susan Schiefelbein. 2007. *The Human, the Orchid, and the Octopus: Exploring and Conserving Our Natural World*. New York: Bloomsbury. https://www.bloomsbury.com/us/human-the-orchid-and-the-octopus-9781596914186/

Cunningham, Sean S., director. 1989. *DeepStar Six*. Carolco Pictures. 99 minutes. https://www.imdb.com/title/tt0097179/

Dickey, Tommy. 1990. "Physical-Optical-Biological Scales Relevant to Recruitment in Large Marine Ecosystems." In *Large Marine Ecosystems: Patterns, Processes, and Yields*. Edited by Kenneth Sherman, Lewis Alexander, and Barry Gold, 82–98. Washington, DC: American Association for the Advancement of the Sciences. http://www.opl.ucsb.edu/tommy/pubs/Dickeyscales1990.pdf

———. 1991. "The Emergence of Concurrent High-Resolution Physical and Bio-Optical Measurements in the Upper Ocean and Their Applications." *Reviews of Geophysics* 29(3): 383–413. https://doi.org/10.1029/91RG00578

———. 2020. "New Discoveries Enabled by the Emergence of High-Resolution, Long-Term Interdisciplinary Ocean Observations." *Perspectives of Earth and Space Scientists* 1(1): 1–10. https://doi.org/10.1029/2020CN000129

Forster, E. S. 1927. *The Works of Aristotle. Volume VII: Problemata*. Edited by W. D. Ross. Oxford: Clarendon Press.

Glen, John, director. 1981. *For Your Eyes Only*. United Artists. 127 minutes. https://www.imdb.com/title/tt0082398/

Hegalson, Nicole. 2016. "First Video of Coral Bioluminescence in the Deep Sea." Reef Builders. August 10, 2016. https://reefbuilders.com/2016/08/10/first-documentation-of-bioluminescence-in-the-deep-sea/

Humphris, S. E., C. R. German, and J. P. Hickey. 2014. "Fifty Years of Deep Ocean Exploration with the DSV Alvin." EOS. June 3, 2014. https://eos.org/features/in-june-2014-the-deep-submergence-vehicle-dsv-alvin-the-worlds-first-deep-diving-sub-marine-dedicated-to-scientific-research-in-the-united-states-celebrated-its-50th-anniversary

Kim, Juliana. 2023. "76 Days and Counting Underwater, Florida Professor Breaks World Record, for Science." NPR. May 16, 2023. https://www.npr.org/2023/05/16/1176131967/underwater-florida-professor-breaks-world-record

Knapp, Alex, and Katie Jennings. 2020. "Fabien Cousteau Is Raising $135 Million to Build the International Space Station of the Deep Sea." *Forbes*. July 22, 2020. https://www.forbes.com/sites/katiejennings/2020/07/22/fabien-cousteau-is-raising-135-million-to-build-the-international-space-station-of-the-deep-sea/?sh=41846a010132

Koch, Kristen. 2019. "California Ecosystem Surveys Give Insight on Ocean Changes to Come." NOAA Fisheries. December 30, 2019. https://www.fisheries.noaa.gov/leadership-message/california-ecosystem-surveys-give-insight-ocean-changes-come

Kreier, Freda. 2022. "Historic Dive Aims to Map Earth's Deepest Point Like Never Before." *Nature*. July 13, 2022. https://doi.org/10.1038/d41586-022-01916-4

Loya, Yossi, Kimberly A. Puglise, Tom C. L. Bridge, eds. 2019. *Mesophotic Coral Ecosystems*. Vol. 12 in *Coral Reefs of the World*. Cham: Springer Nature Switzerland AG. https://doi.org/10.1007/978-3-319-92735-0

Matsen, Brad. 2009. *Jacques Cousteau: The Sea King*. New York: Pantheon Press. https://www.penguinrandomhouse.com/books/109161/jacques-cousteau-by-brad-matsen/

May, Cameron, Lauren Meyer, Sasha Whitmarsh, and Charlie Huveneers. 2019. "Eyes on the Size: Accuracy of Visual Length Estimates of White Sharks, *Carcharodon carcharias*." *Royal Society Open Science* 6: 190456. http://dx.doi.org/10.1098/rsos.190456

Melnick, Kyle. 2019. "NASA Partners with US Navy to Develop AR Displays for Spacesuits." VRScout. July 21, 2019. https://vrscout.com/news/nasa-us-navy-ar-heads-up-display/

Melville, Herman. 1851. *Moby-Dick*. New York: Harper & Brothers.

Mills, Eric. 2012. *Biological Oceanography: An Early History.* 1870–1960. University of Toronto Press. https://utorontopress.com/9781442613720/biological-oceanography/

Morelo, Lauren. 2010. "Reenlisting Submarines to Study Global Warming in the Arctic." *Scientific American*, reprinted from Climatewire with permission from Environment & Energy Publishing. July 20, 2010. https://www.scientificamerican.com/article/re-enlisting-submarines-t0-study-global-warming-in-the-arctic/

Munk, Walter. 2000. "Achievements in Physical Oceanography." In *50 Years of Ocean Discovery: National Science Foundation 1950–2000*. Washington DC: National Academy Press: 44–50. https://doi.org/10.17226/9702

Naval Technology. 2021. "NAVSEA and NASA Partner to Test Generation 2 DAVD System." Naval Technology. March 2, 2021. https://www.naval-technology.com/news/navsea-and-nasa-partner-to-test-generation-2-davd-system/

NOAA. 2023. "NOAA, Proteus Ocean Group to Explore Uses of Groundbreaking Underwater Lab." NOAA website, accessed June 12, 2023. https://www.noaa.gov/news-release/noaa-proteus-ocean-group-to-explore-uses-of-groundbreaking-underwater-lab

Oberhaus, Daniel. 2020. "The Oldest Crewed Deep Sea Submarine Just Got a Big Makeover." *Wired*. December 22, 2020. https://www.wired.com/story/the-oldest-crewed-deep-sea-submarine-just-got-a-big-makeover/

Oreskes, Naomi. 2020. *Science on a Mission: How Military Funding Shaped What We Do and Don't Know about the Ocean*. Chicago: University of Chicago Press. https://press.uchicago.edu/ucp/books/book/chicago/S/bo59258933.html

The Salt Sirens. 2022. "Current List of Freediving Records (2022)." April 27, 2022. https://thesaltsirens.com/current-freediving-records/

SCICEX Science Advisory Committee. 2010 SCICEX *Phase II Science Plan, Part I: Technical Guidance for Planning Science Accommodation Missions*. Arlington, VA: US Arctic Research Commission. https://www.arctic.gov/uploads/assets/scicex_plan.pdf

Thomson, Sir C. Wyville. 1878. *The Voyage of the "Challenger," the Atlantic: A Preliminary Account of the General Results of the Exploring Voyage of the H.M.S. "Challenger" During the Year 1873 and the Early Part of the Year 1876*. New York: Harper and Brothers.

Thomson, Sir C. Wyville, and John Murray. 1885. *Report on the Scientific Results of the Voyage of the H.M.S. Challenger During the Years 1873–76*. Narrative Volume I: First Part. Published by Order of Her Majesty's Government. https://doi.org/10.5962/bhl.title.6513

University-National Oceanographic Labratory System. 2021. *Research Vessel Safety Standards*. 11th ed. Seattle: UNOLS Office. https://www.unols.org/document/research-vessel-safety-standards-rvss

van Haren, Hans. 2018. "Abyssal Plain Hills and Internal Wave Turbulence." *Biogeosciences* 15: 4387–4403. https://doi.org/10.5194/bg-15-4387-2018

Verne, Jules. (1870) 1993. *Twenty Thousand Leagues Under the Sea*. Translated by Walter James Miller and Frederick Paul Walter. Annapolis: Naval Institute Press. https://www.usni.org/press/books/20000-leagues-under-sea

Walsh, Don. 1962. *The Bathyscaph Trieste: Technological and Operational Aspects*, 1958–1961. Report 1096. San Diego: U.S. Navy Electronics Laboratory. https://apps.dtic.mil/sti/pdfs/AD0616288.pdf

Weinman, Steve. 2022. "Deep-Sea Mapper Can't Get Much Deeper!" Divernet. July 15, 2022. https://divernet.com/world-dives/deep-sea-mapper-cant-get-much-deeper/

Wilkins, John. 1708. *The Mathematical and Philosophical works of the Right Reverend John Wilkins, Late Lord Bishop of Chester*. London: Printed for J. Nicholson, A. Bell, B. Tooke, and R. Smith.

Will, Manuel, Andrew W. Kandel, and Nicholas J. Conard. 2019. "Midden or Molehill: The Role of Coastal Adaptations in Human Evolution and Dispersal." *Journal of World Prehistory* 32: 33–72. https://doi.org/10.1007/s10963-018-09127-4

Figure 3.29
Despite the challenges of working at sea, life aboard an oceanographic research vessel can be an exhilarating and unforgettable experience.

Credit: Shaw and Chamberlin

Do robofish dream of electric krill?
Artwork by Martha Rich.

Robots, Satellites, and Observatories

4

Figure 4.1
Modern technology allows oceanographers to explore the world ocean in ways that seemed inconceivable just a few decades ago.

Credit: istock/S_Bachstroem

In Douglas Adams's five-part trilogy *The Hitchhiker's Guide to the Galaxy* (1995), earthling Arthur Dent narrowly escapes Earth's destruction thanks to an extraterrestrial Good Samaritan who is field-testing a not-so-dependable computerized guide to the galaxy. Their adventures, full of strange technological inventions, bear modest similarities to the technology-enhanced lives of oceanographers.

The application of electronic sensors for observing, measuring, and adventuring into the world ocean has provided a view of the ocean as never before (minus the leaps through hyperspace). These sensors—attached to ships, subs, satellites, robots, and even animals—have dramatically changed our ability to investigate and understand the ocean. Our 21st-century view of the ocean feels like a trip to a galaxy far, far away.

The ever-expanding fleet of sensor-outfitted platforms that fly above, glide on top, or dive below the surface of the ocean belongs to a broad category of tools known as **ocean observing systems**. Here we include any platform or sensor that permits observations and measurements at a distance. Independent and self-navigating platforms, such as

Figure 4.2

Oceanographer Carl Lorenzen (1938–1988) on the hero platform aboard the R/V *Thomas G. Thompson* in the 1970s. Oceanography has come a long ways from lowering sampling bottles over the side of a ship.

Credit: Chamberlin

satellites and autonomous underwater vehicles (AUVs), fit the definition to a tee. Tools as simple as drifters and as complex as seafloor observatories may also fall into this category. Some oceanographers might include conductivity-temperature-depth instruments, or CTDs, in this category (covered in Chapter 13). Most importantly, the tools of ocean observing systems share a common goal: to expand the reach of human senses over broad scales of space and time (e.g., Lee et al. 2017).

As oceanographers increase their ability to observe all parts of the ocean and connect measurements across space and time, a new and vital understanding of Earth and its ocean system will emerge. Our ability to survive in the coming centuries may well depend on the robotic eyes and ears of the sea. Like Arthur Dent, we may return to a restored Earth thanks in part to these platforms. That day, when it comes, may bring new meaning to "thanks for all the fish."

Figure 4.3 Until we learn how to communicate with dolphins, we will have to rely on technology to understand how the ocean works as a system. "Siri, what does 'wheeeee click click' mean?"

Credit: istock/Andrea Izotti

4.1 Bottles, Drifters, and Floats

If you've ever seen a plastic bottle adrift on the surface of a lake or the ocean, then you have a pretty good understanding of one of the earliest tools for tracking ocean currents. The **drift bottle method** relies on a return-to-sender card sealed inside a buoyant container—traditionally a glass bottle—that is tossed from a ship or shore. The card requests that the finder return the card with information on the date and location where the bottle was found. By comparing the starting point to the final destination, the path of the drift bottle may be inferred. Drift bottles have served as indicators of surface current speed and direction since at least

Figure 4.4 Cast into the Atlantic Ocean in September 1959, this drift bottle was found on Martha's Vineyard, Massachusetts, in December 2013.

Credit: NOAA's National Ocean Service CC BY 2.0

1802 and remain in use today (e.g., Stommel 1965; Storch et al. 2020). At Woods Hole Oceanographic Institute (WHOI), oceanographer Dean "Bump" Bumpus (1912–2002) oversaw the release of hundreds of thousands of drift bottles during his 40-year career (Lipsett 2014; Woods Hole 2021).

While drift bottles are not autonomous, their use inspired two types of passive float technologies in use today: drifters and floats. **Drifters** (a.k.a. surface drifting buoys) provide information on surface currents, and **floats** track currents at deeper depths. As WHOI physical oceanographer Bruce Warren (1937–2010) puts it, "Drifters float and floats sink" (e.g., Lumpkin et al. 2017). Drifters and floats represent what are known as **Lagrangian platforms**—devices that move with the currents—named after Italian mathematician Joseph-Louis Lagrange (1736–1813), who developed the mathematics of fluid flows (e.g., Knauss and Garfield 2017).

In the 1950s English oceanographer John Swallow (1923–1994) developed a float that could maintain a constant depth and transmit an

Figure 4.5 Lagrangian path of a Swallow float at around 600 meters in the eastern Atlantic Ocean (Swallow 1955).

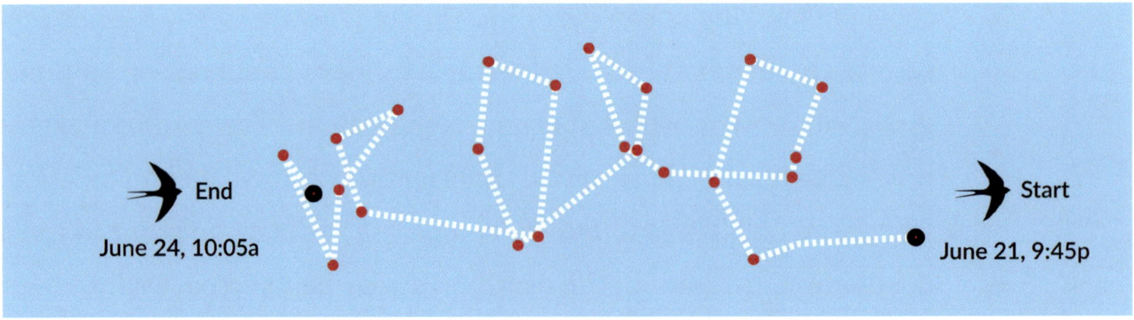

Credit: Chamberlin

Figure 4.6 Oceanographer John Swallow—under the watchful eye of the ship's cat—prepares a float for deployment aboard *Discovery II* in 1957.

Credit: Courtesy of National Oceanographic Library, Archives, National Oceanography Centre, Southampton, UK

acoustic signal so that it could be tracked by a ship or a receiving station (Swallow 1955). These acoustically transmitting floats—known as **Swallow floats**—provided some of the first hints at the complex, eddy-like nature of the ocean (e.g., Crease 1962).

In 1979 the National Oceanic and Atmospheric Administration (NOAA), along with the World Meteorological Organization and Intergovernmental Oceanographic Commission, established the Global Drifter Program (GDP; Niiler 2001). The most common drifter in use today, the Surface Velocity Program (SVP) drifter, consists of a small, spherical surface float attached to a nearly 23-foot-long (7 m) cloth cylinder with holes in it (to reduce drag), the so-called **holey sock drogue**. As of August 2022, some 1,244 drifters have been deployed by 26 countries. Each transmits position and temperature data to a satellite for upload to NOAA's GDP database and website.

Swallow floats and SVP drifters set the stage for development of **autonomous profiling floats**, underwater robots that collect, store, and

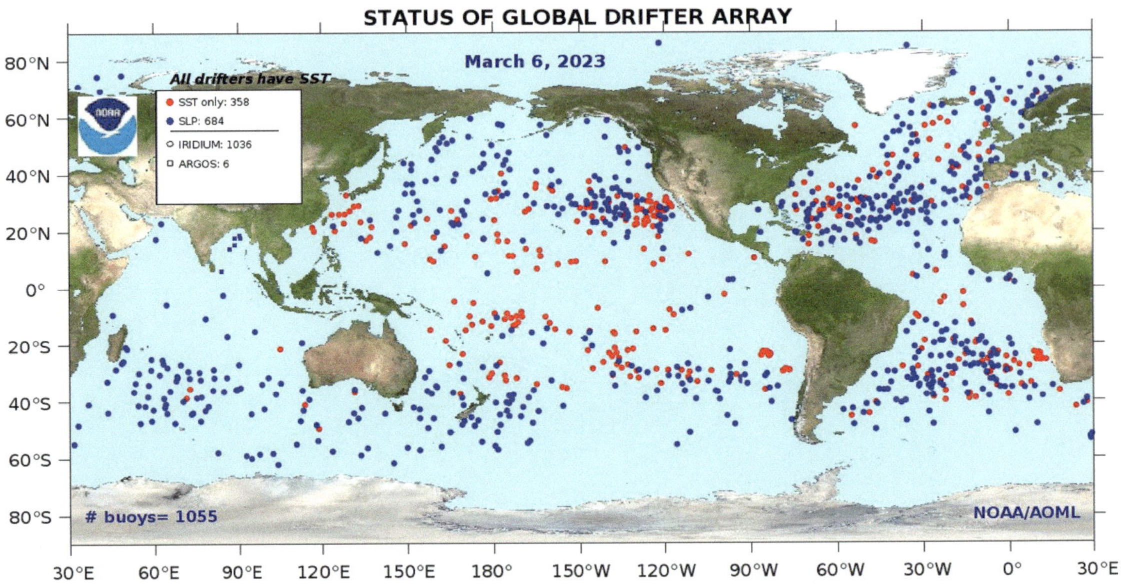

Figure 4.7
Status of the Global Drifter Array as of March 6, 2023.

Credit: NOAA/Atlantic Oceanographic Meteorological Laboratory

report data via satellite. The crown jewel of autonomous profiling floats is the Array for Real-Time Geostrophic Oceanography, known simply as Argo.

Put into service in 1999, the Argo floats take their name from the legendary ship *Argo*. Commanded by the Greek hero Jason, *Argo*, with his heroic crew, the Argonauts, set sail to find the Golden Fleece, the fur of a winged golden ram, as told in Greek mythology (Colavito 2014). The Argo floats have been hailed as "one of the scientific triumphs of our age" (Gillis 2014). Fittingly, the Argo floats serve under the watchful eye of the ocean-observing Jason satellites, named for the Greek mythological hero first mentioned by Homer in 800 BCE (Colavito 2014). Together they provide simultaneous ocean measurements above and below the surface.

Figure 4.8
Modern replica of the legendary ship *Argo* for which a fleet of autonomous profiling floats are named.

Credit: istock/ Jekaterina Voronina

An individual Argo float resembles a tall and skinny version of R2-D2, the feisty robot in the Star Wars movies. They move up and down in the water column by regulating their **buoyancy**, the balance between rising or sinking in the water column (see Chapter 13). As anyone who has used a floatie knows, your ability to float on the surface depends on the volume of air inside the floatie. Let the air out and you sink. Fill the float with air and you float. An Argo float works in a similar fashion, except instead of air, the float uses oil in an internal reservoir. When pumped into a flexible bladder at the bottom of the float, the oil occupies a greater volume and makes the float more buoyant. When the oil is removed, the bladder shrinks and the float sinks. By carefully controlling the bladder, the float can sink, rise, or stay at a constant depth.

Credit: NOAA/Pacific Marine Environmental Laboratory

Figure 4.9
An army of Argo floats ready for deployment.

Argo floats can measure, store, and transmit observations on a number of physical, chemical, and biological properties, including temperature, salinity, pH, oxygen, and nitrate. As of August 2022, a flotilla of 3,930 floats report data from the surface to depths of 6,562 feet (2,000 m) across the entire

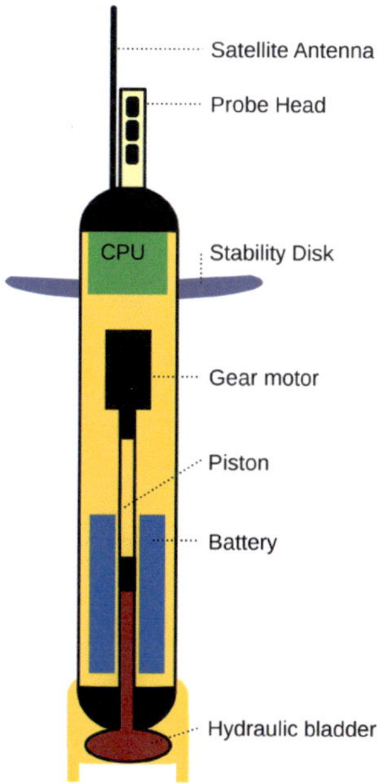

Credit: Wikipedia/Brn-Bld CC BY-SA 3.0

Figure 4.10
Anatomy of an Argo float. The piston controls buoyancy. The satellite antenna transmits data when the buoy comes to the surface.

Figure 4.11 Argo floats operate on a 10-day cycle. GPS positions at the start and end of the cycle provide data on their movement in the ocean.

Credit: Shaw and Chamberlin

world ocean, including the Southern and Arctic Oceans. Over the span of their more-than-two-decade lifetime, the Argo floats have generated more than two million profiles of the upper ocean, nearly four times the number of profiles available prior to their implementation. More than 2,000 scientific papers have used Argo data to support their findings (e.g., Jayne et al. 2017).

In recent years a new generation of Argo floats—Deep Argo—are being deployed to obtain data over the entirety of the world ocean. Argo floats with nicknames such as Deep SOLO and Deep APEX developed in the United States will reach depths up to 19,685 feet (6,000 m). Deep NINJA and Deep ARVOR, under development by Japan and France, respectively, will operate at depths to 13,123 feet (4,000 m; Jayne et al. 2017). Oceanographers have also developed a new suite of sensors for measurements of optical and biogeochemical properties, a fleet of floats called Biogeochemical Argo

Figure 4.12 Deployment of an Argo float from a ship involves switching it on and tossing it over the side.

Credit: NOAA

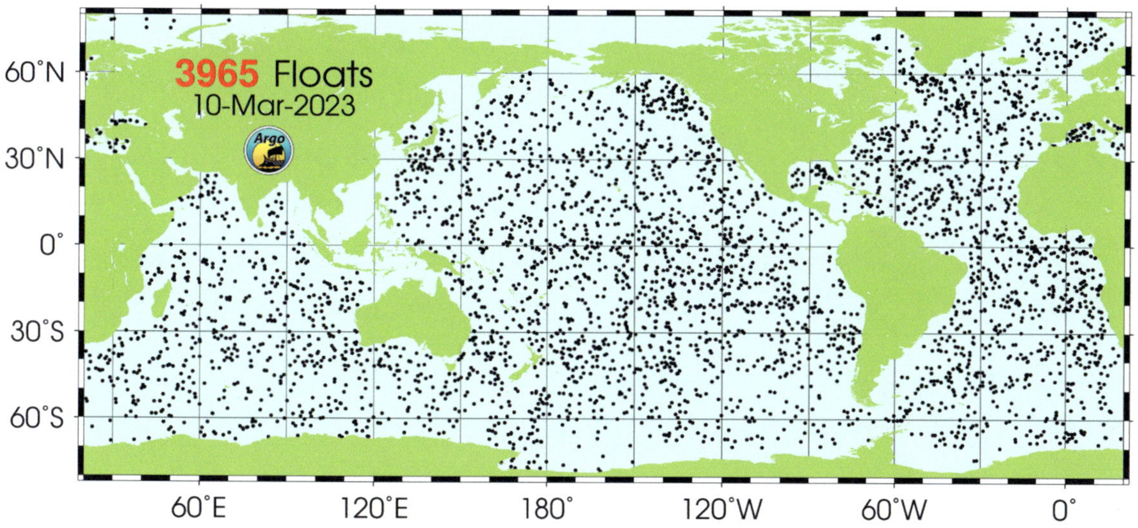

Figure 4.13
As of March 10, 2023, nearly 4,000 Argo floats are providing near real-time data on the upper ocean (<2,000 m).

Credit: Argo Program

(e.g., D'Ortenzio et al. 2020). These new Argo floats promise to shed light on the role of the deep ocean in climate change and improve our understanding of its physics, chemistry, and biology (Roemmich et al. 2019).

While all the above drifters and floats are traditionally deployed from ships, aircraft can launch floats too. The Air-Launched Autonomous Micro-Observer (ALAMO) can profile the ocean to depths of more than 3,900 feet (1,200 m). ALAMO floats were developed in response to a need for obtaining observations in places difficult to reach, such as ice-covered waters and the interior of hurricanes (e.g., Jayne and Bogue 2017; Goni et al. 2017). Deployments of ALAMO floats during the 2017 and 2018 Atlantic hurricane seasons yielded new insights into the response of the upper ocean to the passage of a hurricane and provided

Figure 4.14
Deep SOLO deployment off the coast of Hawaii.

Credit: NOAA/PMEL

Figure 4.15
ALAMO floats are deployed from aircraft and float to the ocean surface on a parachute.

Credit: NOAA/PMEL

ground-truth data for hurricane forecast models (Sanabia and Jayne 2020). Ultimately, the better our understanding of factors that influence hurricane intensity, the more likely hurricane forecasts can provide accurate warnings and save property and human lives.

4.2 Gliders

Torpedo-shaped, with wings and a tail rudder, **gliders** "fly" through the water column like an underwater airplane. But gliders have no propellers or engines. Instead, gliders rely on buoyancy changes to propel them through the water. When they lose buoyancy, the nose of the glider becomes heavier and it sinks. But instead of sinking vertically, its wings cause it to move forward at an angle. Similarly, as the glider gains buoyancy and the nose lifts, the wings cause it to move forward in the water column at an angle. The resultant flight pattern resembles a sawtooth. While slow moving relative to propelled vehicles, the efficiency of their movement and their low cost allow them to carry out missions of a month or more (e.g., Schofield et al. 2007).

Figure 4.16
Gliders sink and rise through the upper ocean collecting valuable spatial information on ocean properties.

Credit: Shaw and Chamberlin

Many types of gliders exist, but the best known is the Slocum glider, named after Canadian-American adventurer

Joshua Slocum (1844–1909), the first person to sail alone around the world (1895–1898). American engineer Douglas Webb (*b.* 1929) originally conceived of the glider in 1978 and then jointly developed it in the 1980s and 1990s with American oceanographer Henry Stommel (1920–1992; Nehring 2021). Like its namesake, the Slocum glider evokes a spirit of exploration and determination—but for robots.

Figure 4.17 This glider operated by Oregon State University was outfitted with hydrophones to record the West Mata underwater volcano off Samoa.

Credit: Oregon State University/ Bob Dziak CC BY-SA 2.0

The latest generation of gliders—**wave gliders**—use the energy of ocean waves for propulsion. The above-water portion of the platform—the float—resembles a surfboard, albeit one outfitted with an antenna and solar panel. Below the surface, a tether connects to an underwater sled—the sub—which features multiple fins arranged like Venetian blinds. These fins generate thrust when the float rises and falls in the ocean swell. Both float and sub may be outfitted with sensors (e.g., Pagniello et al. 2019; Grare et al. 2021). In 2013 a wave glider named Benjamin Franklin completed a 455-day journey of 7,939 nautical miles (14,703 km) from San Francisco Bay to Lady Musgrave Island off the coast of Australia. The feat earned the platform a Guinness World Record for the "longest journey of an unmanned autonomous surface vehicle" (Guinness World Records 2013).

Figure 4.18 A wave glider uses the power of waves and solar energy to travel thousands of miles across the ocean. This one was photographed off the coast of Hawaii.

Credit: Stephen Chin CC BY-SA 2.0

Longer missions and faster coverage make gliders an ideal platform for oceanographic research in remote areas of the world ocean rarely visited by ships.

4.3 Towfish

Originally deployed in the 1960s in sonar systems for mapping the seafloor and locating sunken craft, the **towfish**, a winged platform pulled by a cable behind a moving vessel, remains an important tool in modern oceanographic research. The platform's resemblance to a torpedo and submariners' use of "fish" as a slang for torpedoes (e.g., Newpower 2006) likely led to adoption of the term for oceanographic purposes (e.g., Flemming 1976).

The simplest towfish resemble winged torpedoes. Other towfish look like sleds with wings attached. The wings allow the vehicles to "fly" up, down, or sideways while carrying instruments to measure various ocean properties. A pump may even be attached to deliver water samples to the surface. Some towfish carry nets to collect plankton, the microscopic organisms that drift with the currents in the ocean. Remotely operated towfish vehicles feature automated control functions. The vehicle can maintain a fixed depth or fly in what oceanographers refer to as the **tow-yo pattern**, a combination of towing and yo-yoing. Geologists were

Figure 4.19 Towfish rely on a moving vessel for propulsion and power, but carry bigger payloads, provide real-time data, and take water samples.

Credit: Shaw and Chamberlin

Figure 4.20 The Australian government used the *Dragon Prince* deep towfish to search for wreckage of Flight MH370, which mysteriously disappeared on March 8, 2014.

Credit: Australian Transportation Safety Board CC BY 4.0

among the first to deploy towfish as part of side-scan sonar systems, which use sound pulses directed toward the seafloor at low angles to produce high-resolution images. Side-scan sonar has been instrumental in locating downed aircraft and sunken ships and providing highly detailed views of the seafloor (e.g., Schlee et al. 1995; Wu et al. 2021).

4.4 Remotely Operated Vehicles

This next piece of technology has become so popular that you can find it for sale on Amazon. **Remotely operated vehicles** (ROVs) owe their inspiration to the human desire to find shipwrecks. While we commonly think of ROVs as a new technology, their origins date back to the 1950s, when French diver-photographer Dimitri Rebikoff (1921–1997) launched the first ROV, the *Poodle*. On its first mission in 1954, the ROV *Poodle* found ancient Phoenician shipwrecks more than 540 feet (165 m) deep (Marx 1990).

Of course, the most famous recent ROV discovery came in 2023 when the American ROV *Odysseus* located the remains of the tourist submersible, *Titan*. On June 18, 2023, the *Titan* imploded with five passengers aboard on a visit to the passenger ship *Titanic* at 12,500 feet deep (3.8 km) in the North Atlantic (Mongilio 2023). *Titanic*, as you know,

Figure 4.21 French inventor Dimitri Rebikoff poses with his many underwater inventions, including his *Poodle* (front), the first ROV ever built.

Credit: Rebikoff Foundation

collided with an iceberg in 1912 and sank, killing 1,500 people. More than 70 years later, an American-French expedition under the guidance of WHOI oceanographer Robert Ballard located the vessel using an ROV (e.g., Ballard and Drew 2021).

Modern ROVs serve a broad range of military, industrial, scientific, and commercial needs. They come in a variety of sizes, shapes, and capabilities (e.g., Moore et al. 2010). The smallest ones are barely the size of a toaster, while the largest ones are bigger than a compact SUV. Most "swim" using a system of propellers controlled via commands sent through a fiber-optic cable called a **tether** (Christ and Wernli 2014). The ROV operator, the pilot, may use a joystick on a control console to maneuver the ROV. If you're good with video games, you would probably make a good ROV pilot, especially if you know a little oceanography (hint, hint).

Some ROVs have wheels or treads like a tank that permit them to crawl over the seafloor, but most resemble snowmobiles, with sleds stretching the length of the vehicle so that they can land gently on the seafloor. Depending on

Figure 4.22 ROVs provide the ability to work underwater like few other craft. Here an ROV explores a sunken barge in Thunder Bay National Marine Sanctuary in Lake Huron.

Credit: NOAA/Tane Casserley

the application, an ROV may be outfitted with one or more **manipulator arms**, extendable, rotational appendages that accommodate a variety of specialized attachments. Grippers hold objects and grabbers can pick them up. Shovels can be attached for digging, tubes deployed for collecting sediments, suction devices used to collect animals, and any number of other devices for sampling and carrying out research (Sivčev et al. 2018). It's quite the thrill to watch an ROV collect strange creatures never before seen by humans.

Credit: NOAA/Jeremy Potter CC BY 2.0

Figure 4.23
During the Hidden Ocean Arctic 2005 Expedition off Point Barrow, Alaska, the ROV *Global Explorer* collected zooplankton using remotely controlled sampling chambers at depths of more than five miles (9,000 m).

Many ROVs are equipped with cameras and banks of lights for high-definition photography and videography. Some even stream live over the internet. Oceanographers refer to this sensory technology as **telepresence**, a kind of virtual reality that lets viewers experience a place that is otherwise inaccessible to (most) humans. Telepresence allows land-based scientists (and you) to participate in missions, too. Experts from around the world watch the livestream and call in to offer scientific expertise and commentary on any of the features or organisms that the ROV encounters. Livestreams

Credit: GEOMAR CC BY 4.0

Figure 4.24
The manipulator arm of the ROV *Kiel 6000* collects a deep-sea sponge attatched to a manganese nodule at about 2.5 miles deep (4,100 m) in the Peru Basin.

Figure 4.25 HD video camera operator Roland Brian brings the seafloor into focus during a dive of the ROV *Deep Discoverer* on the 2016 Deepwater Exploration of the Marianas Expedition.

***Credit*: NOAA**

offer informative and occasionally goofy commentary on the geology, chemistry, physics, and biology of the seafloor. I confess to standing in line at the supermarket watching the live feed of an ROV camera probing the ocean depths from locations halfway across the world. Check it out next time they go live. (See https://oceanexplorer.noaa.gov/.)

4.5 Autonomous Underwater Vehicles

If ROVs had besties, they would be **autonomous underwater vehicles**, better known as AUVs, untethered craft that operate independently on pre-programmed missions. In fact, the US Navy lumps ROVs and AUVs into a single category: **uncrewed underwater vehicles** (UUVs). AUVs represent the ultimate expression of a robotic deep-sea explorer. Free of any cables, commanded by a computer brain, and potentially able to wander the ocean for unlimited periods of time, AUVs travel in a way that ROVs cannot. Most fly through the water column. Some crawl along the seafloor. Most impressive is their ability to maintain level flight and react to circumstances with limited decision-making capabilities.

Though robotic solo flights present a number of technological and engineering challenges, AUVs have begun to make their mark on 21st-

Figure 4.26
The AUV *Abyss*, whose nickname is Tiffy, can reach depths of 3.7 miles (6,000 m). In 2018, it was used for high-resolution mapping of the seafloor in the Lau Basin near the Kingdom of Tonga.

Credit: **GEOMAR/*Philipp Brandl* CC BY 4.0**

century oceanographic research (e.g., Searle et al. 2018; Meyer et al. 2019). In a 2015 study off Guadalupe Island on the west coast of Baja California, researchers used AUVs to track the behavior of great white sharks (*Carcharodon carcharias*). Though the sharks regularly attacked the AUVs (they apparently don't like being followed by a yellow torpedo), the work yielded new insights into the behavior of these magnificent animals (Kukulya et al. 2016).

Researchers have also begun to use AUVs with **adaptive tracking systems**. In essence, the AUV detects and automatically tracks specific features, such as a particular temperature of water or concentration of plankton. Researchers working off Hawaii in 2018 employed two long-range AUVs to automatically track and sample an **ocean eddy**—a self-sustaining rotating loop of water tens to hundreds of miles wide. Eddies can generate conditions ideal for the growth of **phytoplankton**—microscopic plant drifters—which concentrate in a layer known as

Figure 4.27
AUV observations of marine animals are providing new insights into their behavior.

Credit: istock/vladoskan

the deep chlorophyll maximum. By programming the AUVs to track the chlorophyll maximum, researchers were able to observe this subsurface feature in unprecedented detail. To an oceanographer, that's impressive (Zhang et al. 2019; Zhang et al. 2021).

The latest generation of AUVs promises to greatly expand their capabilities. Hovering AUVs maintain their position over a phenomenon of interest such as a hydrothermal vent. Stealth AUVs rest on the seafloor listening for military targets. New, silicone-based soft robots—so-called **bionic AUVs**—mimic animals to blend in and observe organisms less intrusively. **Uncrewed surface vessels** (USVs) allow scientists (and the military) to make observations in remote locations (Braginsky et al. 2020). A catamaran-style USV can even tow a plankton net for sampling microplastics, plankton, and other floating particles (Bazilchuk 2022). For young ocean scientists and engineers, AUVs represent an emerging opportunity to push the boundaries of ocean exploration, a chance to go where no oceanographer has gone before (e.g., Katzschmann et al. 2018; Li et al. 2021; Mazzolai et al. 2022).

Figure 4.28
The US Coast Guard recently tested the capabilities of uncrewed surface vessels for surveillance in remote locations.

Credit: US Coast Guard

4.6 Hybrid Platforms

Hybrid platforms combine the best features of ROVs and AUVs. Formally known as **hybrid remotely operated vehicles** (HROVs), these platforms permit researchers to operate the same vehicle with or without a tether (e.g., Bowen et al. 2009; Yoerger et al. 2021). As outlined by Xiang et al. (2015), an HROV can launch underwater in ROV mode and be guided to a research site, then operate as an AUV and return to the surface when its mission is complete.

Hybrid platforms allow oceanographers to reach the deepest depths in the ocean. The Japan Agency for Marine-Earth Science and Technology, or JAMSTEC, developed the UROV11K—with 12.4 miles (20 km) of fiber-optic cable. In 2017 the vehicle reached a maximum depth of 35,758 feet (10,899 m) in the Mariana Trench (Nakajoh et al. 2018). Unfortunately, the UROV11K has since sunk (Wang et al. 2020). HROVs remain a specialized robotic platform, yet their ability to receive and transmit data and move more flexibly and at greater distances than ROVs will likely propel their further development as a tool for undersea exploration and research.

Figure 4.29
The HROV *Ariane* developed by the French Research Institute for Exploitation of the Sea (IFREMER) can operate at depths up to 8,202 feet (2,500 m) for seafloor mapping and sampling of the seafloor.

Credit: iFremer/Dugornay Olivier CC BY 4.0

4.7 Benthic Landers

Not all robotic craft operate in the water column. A group of platforms known as **benthic landers**, nicknamed ocean elevators, sink to the seafloor. Once landed, they carry out any number of observations, such as imaging, remote sensing of ocean properties, sample collection, or experimentation. Landers have proved particularly useful for studying submarine canyons, where episodic events such as underwater mudslides occur. They're also perfect for observations in deep oceanic trenches that are largely inaccessible to other craft. The deepest-diving landers, the **hadal landers**, deploy baited cameras and traps to attract, film, and even capture deep-sea organisms. In 2014 oceanographers used landers to catch a **snailfish**, the deepest known vertebrate, collected at 26,135 feet (7,966 m) in the Mariana Trench (Gerringer et al. 2017). In 2017 a JAMSTEC lander filmed a snailfish at 26,831 feet (8,178 m), the deepest observation of any fish to date (JAMSTEC 2017). Hadal landers have also been used to study sediments and oxygen consumption in oceanic trenches (Lumberg et al. 2018; Glud et al. 2021). Though less sexy than propelled vehicles, landers offer one of the most cost-effective means for making observations in the deep ocean (e.g., Giddens et al. 2021; Du et al. 2023).

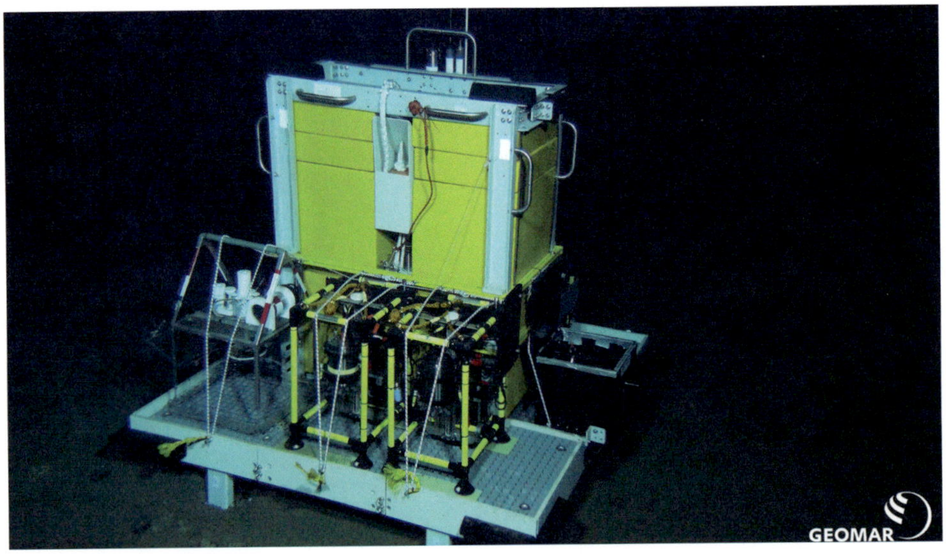

Figure 4.30 A benthic lander carries out respiration experiments and collects specimens at the DISCOL deep-sea disturbance site in the Peru Basin.

Credit: GEOMAR/ROV-Team CC BY 4.0

4.8 Mini-Autonomous Underwater Explorers

Sometimes smaller is better. Enter the **mini-autonomous underwater explorer** (M-AUE). The M-AUE system operates as a swarm of instruments, a dense collection of subsurface drifters and GPS-enabled surface floats that track upper ocean motions in a kind of virtual electronic net. Roughly the size of stout 1.5-liter sports bottles, the subsurface drifters can provide a three-dimensional view of the upper ocean over distances of hundreds of feet to a few miles. Because the drifters can be programmed to automatically maintain their position in the water column, they can mimic the drift patterns of plankton, harmful algae, or oil spills (e.g., Jaffe et al. 2017). The fine-scale information provided by M-AUEs has already revealed insights into how marine larvae find their way home (e.g., Garwood et al. 2020). M-AUEs offer a preview of an emerging field of study known as **underwater swarm robotics**, the design and development of multi-robot systems that interact and coordinate their behavior to solve problems and carry out complex tasks (e.g., Connor et al. 2021). Such systems offer opportunities to better understand the connections between physical and biological processes in the ocean (e.g., Dickey 1991).

Figure 4.31 **Miniaturized platforms permit deployment of swarms of instruments to obtain a 3D view of the ocean's properties.**

Credit: Shaw and Chamberlin

4.9 Ocean Moorings

Ocean moorings provide a fixed platform to which various oceanographic instruments can be attached to acquire measurements of ocean properties for long periods of time. They excel in sampling temporal scales at an unprecedented resolution—minutes to weeks—and they can be left unattended for months to years at a time.

The foundation of an ocean mooring involves a piece of machinery all too familiar to us. Next time you're stuck at a railroad crossing, check out the wheels on the train. At several hundred pounds each, they make great anchors. And indeed, that's what oceanographers use them for. (And just to clear this up, there's absolutely no truth to the rumor that oceanographers steal wheels from parked trains at night.) Train wheels provide the foundation for a steel cable to which floats (often hollow glass balls) and various oceanographic instruments can be attached. The cable may stretch vertically all the way from the seafloor to the surface, or it may rise from the seafloor to a particular depth of interest. Meteorological instruments attached to the surface floats collect weather data. Below the surface, oceanographic instruments may be attached at various intervals to collect depth-specific information.

While several types of instruments may operate simultaneously at fixed depths, profiler moorings allow instruments to travel up and down a cable and sample the water column in a manner similar to an instrument on a cable on a ship. In either case—fixed or profiling—when an oceanographer wants to retrieve the instruments, they send a secretly coded acoustic signal to the mooring

Figure 4.32
Oceanographers have used train wheels as anchors for ocean moorings since at least the 1960s.

Credit: istock.com/Artur Nichiporenko

Figure 4.33 Ocean moorings enable sampling of water column properties over very long timescales and play a critical role in satellite ground truthing.

Credit: NOAA/PMEL

that activates a release mechanism, unhooking the cable from the railroad wheel and permitting the instruments to float to the surface. The wheel remains on the seafloor, forever a part of Davy Jones's locker.

Moorings have provided some of the longest continuous observations of the ocean on record. Off the coast of Bermuda in the Sargasso Sea (or the Bermuda Triangle, as some people call it), oceanographers with the Ocean Flux Program (OFP) have maintained a mooring since 1978. The OFP mooring hosts **sediment traps**, devices that collect and store sinking particles. Its data—more than 40 years' worth—offers invaluable insights on the ocean's recent past. These data have been especially useful for assessing the effects of climate change in the ocean (e.g., Conte and Weber 2014; Cael et al. 2021).

Because ocean moorings can provide critical meteorological information, the US established in 1967 the National Data Buoy Center (NDBC). The NDBC oversees the operations and data quality control of hundreds of moored buoys deployed in US coastal and oceanic waters across the Atlantic and Pacific. It also provides quality control and dissemination of data from hundreds of partner stations. Its website

Figure 4.34
The Central Surface Mooring on the Pioneer Array off the New England coast. The buoy supports surface instruments and maintains the anchored instrtument packages in a near-vertical position. The anchor may also serve as a platform for instruments.

Credit: Image from work supported by the Ocean Observatories Initiative, a major facility fully funded by the National Science Foundation (NSF).

provides near-real-time data from stations in the Atlantic, Pacific, and Indian Oceans. The NDBC serves a wide range of scientific, military, governmental, private, and civilian needs. Fishers, sailors, and boaters, among others, go there for up-to-date information about the waters in which they travel.

Figure 4.35 The Deep-ocean Assessment and Reporting of Tsunami (DART) buoys provide an early warning system for potential tsunami hazards in coastal regions.

Credit: NOAA/Pacific Marine Environmental Laboratory

Despite their many advantages, the harsh conditions to which moorings are subjected can cause breakdowns, corrosion, and, on occasion, total loss. Moorings attract fish that bite and fray the cables. Fish attract fishers, who accidentally snag moorings when trawling or intentionally latch onto moorings and steal their parts. A cable snaps or the anchor accidentally releases, and the mooring drifts away, never to be seen again. On some occasions, the release fails to respond to its acoustic signal, in which case the crew must drag an anchor through the water hoping to catch the subsurface cable and bring the mooring to the surface. Rescue missions to find moorings take up costly ship time and fray the nerves of the oceanographers responsible for maintaining them. Nonetheless, the benefits they serve, such as El Niño prediction, tsunami warning, storm warning, and weather forecasting, more than outweigh their costs. We would be blind to what's happening beneath the ocean's surface without them.

Figure 4.36 Fish love moorings, but they attract fishers, who may accidentally damage cables and instruments.

Credit: istock/ Manjun Shen

Figure 4.37
New technologies have exapnded oceanographers' ability to sample the ocean over broad scales of space and time. By integrating different platforms—each with their own strengths and weaknesses—oceanographers can overcome many of the sampling challenges of the past. The future of oceanography has never been more promising.

nexos

4.10 Uncrewed Aerial Platforms

Figure 4.38
NOAA scientists are using hexacopters to track whales and sample their breath.

Credit: NOAA

Increasingly, scientists are turning to **uncrewed aerial platforms**—drones—remotely controlled flight-capable platforms outfitted with video cameras and various sensors. This category of platforms also includes remotely piloted aerial systems and multirotor copters (e.g., Klemas 2015; Nguyen et al. 2022). The use of drones and drone-based sensors in oceanographic research represents one of the most recent examples of how new technological developments can revolutionize a field (e.g., Gray et al. 2022).

One of the earliest uses of drones for oceanographic research was simply to follow whales and dolphins (Fiori et al. 2017). Images and video from drones have enabled scientists to map intertidal reefs (e.g., Murfitt et al. 2017), track discharge plumes of groundwater (Lee et al. 2016), survey the size and shape of beaches (Casella et al. 2020), and measure snow thickness on sea ice (Molar-Candanosa 2022).

Figure 4.39
In this concept image, drones with hyperspectral and thermal sensors may one day help map wildfires.

Credit: NASA/Ames Research Center/Daniel Rutter

A new generation of sensors—deployable on drones and aircraft—now permit oceanographers to measure a wide variety of ocean properties. **Multispectral imagers** detect several different wide bands of electromagnetic radiation (colors). **Hyperspectral imagers** detect hundreds of narrow bands of electromagnetic radiation. Both routinely fly aboard airborne platforms to monitor the environment (Stuart et al. 2019). Using a drone-based hyperspectral sensor, researchers were able to document the extent and concentration of a harmful algal bloom in two bays in southern China (Li et al. 2021).

Figure 4.40
The USGS is using camera-equipped kite-balloons to map beaches.

Credit: **USGS/Sandra Brosnahan**

Not all modern oceanographic measurements require cutting-edge platforms. Kites, outfitted with state-of-the-art cameras or sensors, serve many coastal applications. **Kite aerial photography** has been used to map the distribution of organisms in the rocky intertidal and near-coastal zone and to acquire high-resolution images of entire islands or beaches (Bryson et al. 2013).

Balloons play a role in oceanographic research too. During the 2018 Microbiology-Ocean-Cloud-Coupling in the High Arctic (MOCCHA) campaign, oceanographers deployed a tethered balloon to sample biologically generated aerosols and the properties of low clouds (Zinke et al. 2021). In the near future, scientists hope to deploy **steerable**

Figure 4.41
NASA is deploying super-pressure, high-altitude balloons to study Earth's atmosphere, electromagnetic field, and the cosmos.

Credit: **NASA**

Figure 4.42 A saildrone uses wind and solar power to supply energy for long-term deployments of ocean sensors.

Credit: US Coast Guard/Tori Barrett

stratospheric balloons, such as Stratollites. As big as a blue whale, these craft can maintain a position above Earth's surface for several days (Good et al. 2018). Stratollites may be used to monitor storms, assist natural disaster efforts, and even remotely sense the ocean. By 2024 you'll be able to ride 19 miles high above Earth in a Stratollite (World View 2023).

And while technically not an aerial platform, the newest tool for carrying oceanographic instruments over large swaths of the ocean involves a hybrid device called a **saildrone**. Part windsurfer, part oceanographic buoy, this uncrewed wind- and solar-powered device can stay at sea for months at a time. On a 60-day voyage from San Francisco to Guadalupe Island, a saildrone carrying 16 ocean and atmospheric sensors provided unprecedented data on key oceanographic and atmospheric processes (Gentemann et al. 2020). NOAA soon hopes to deploy fleets of saildrones to replace some of its aging buoys at a fraction of the cost (Voosen 2018). Of course, on windless days, saildrones might also make a nice resting spot for migrating seabirds or a peaceful place for a textbook-weary student to get away from it all.

Sometimes the best way to observe the ocean is above its surface, where you can get a bird's-eye view.

4.11 Satellite Oceanography

Space-based observations of the ocean have revolutionized oceanography. Though physical oceanographers had long suspected the chaotic nature of ocean circulation, the first satellite images of ocean color in 1978 drew a collective "oh, my." The surface of the ocean and the phytoplankton suspended within it resembled the sky in Vincent van Gogh's painting *Starry Night* (1889): variable, complex, and organized in loose, rotating structures. The ocean's **mesoscale structure**—the highly variable eddies that characterize nearly all of the ocean's flows—suddenly became visible. Seemingly overnight, **satellite oceanography**, the study of the world ocean using sensors aboard space-based platforms, changed the way we thought about and studied the ocean. Barber and Hilting (2000) likened it to "seeing the ocean for the first time."

Prior to this grand awakening, weather and military satellites had routinely observed the Earth, ice caps, and ocean. But on June 26, 1978, the National Aeronautics and Space Administration (NASA) launched the very first satellite dedicated solely to oceanography, the Seafaring

Figure 4.43 The GOES-18 satellite, launched on March 1, 2022, represents the most advanced weather satellite to date for observing Earth's atmosphere and surface.

Credit: NOAA/United Launch Alliance

Figure 4.44
Painted in 1889, Van Gogh's *Starry Night* resembles the color imagery of ocean satellite images a century later.

Credit: Wikipedia (left); NASA (right)

Satellite, or Seasat. Four months later, NASA launched the Nimbus 7, which carried the Coastal Zone Color Scanner (CZCS), the first sensor designed to measure ocean color. Its success drew greater attention to the optical properties below the ocean's surface and gave birth to the field of **bio-optical oceanography**, the study of the submarine light field and its variations due to biological processes (Smith and Baker 1978). This extraordinary technology continues to make significant contributions to our understanding of the world ocean in the 21st century.

Figure 4.45
Researchers in the Great Lakes carry out measurements of underwater light using a radiometer, a critical tool for verifying the data obtained by satellites.

Credit: NOAA Great Lakes Environmental Research Lab CC BY-SA 2.0

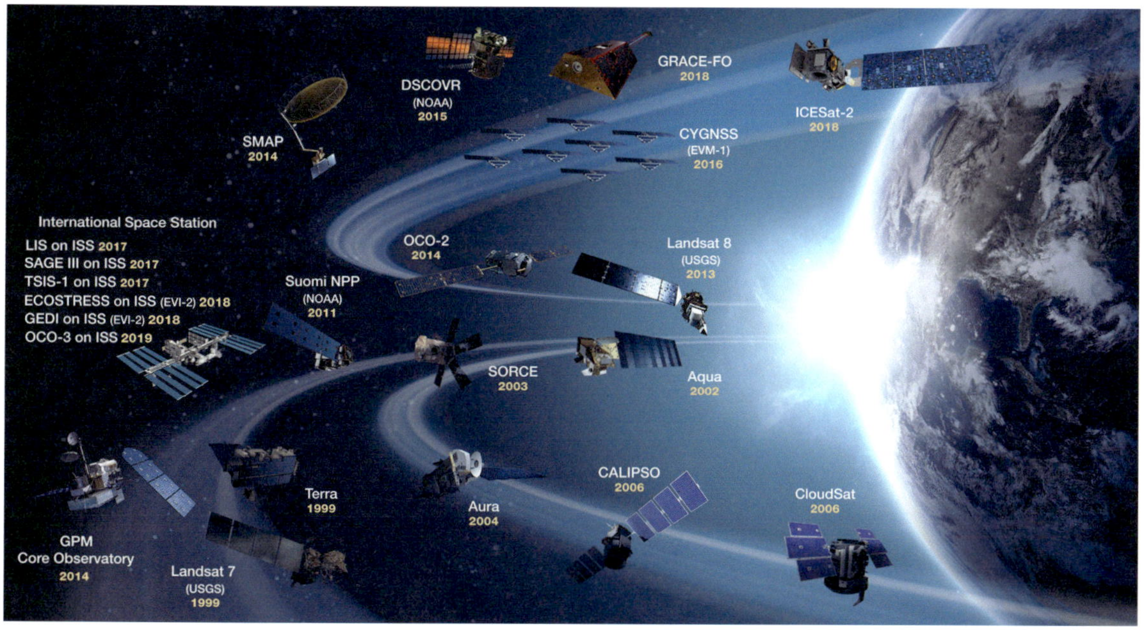

Figure 4.46 The current fleet of NASA Earth-observing satellites and sensors flown aboard the International Space Station continuously monitor Earth system processes.

Credit: NASA

4.11.1 Sea Surface Temperature

Covering 71 percent of Earth's surface, the world ocean dominates temperature regulation on our planet. As NOAA puts it, "the ocean is the largest solar energy collector on Earth" (Lindsey and Dahlman 2020). Through heat exchange across the **air-sea interface**—the boundary between the ocean and the atmosphere—the world ocean adds or removes heat from the atmosphere and vice versa. These exchanges of heat drive the motions of the atmosphere and the ocean and prove important for a whole heckuva lot of other things we'll learn about in the chapters ahead.

Satellite oceanography owes much of its early development to sensors used in aerial oceanography. One group of sensors, collectively known as **radiometers**, detects different kinds of electromagnetic radiation, the energy of light. Because the ocean and the stuff it contains remove,

Space-based platforms have revolutionized the way we view and understand the world ocean.

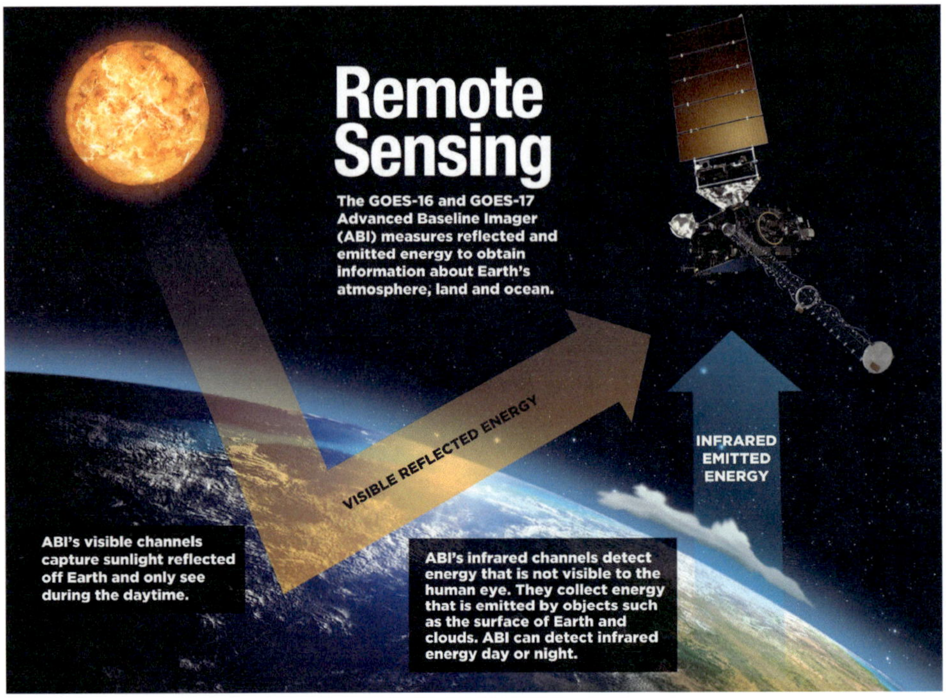

Figure 4.47 Remote sensing of Earth's surface relies on sensors that detect different bands of electromagnetic radiation, that is, light of various wavelengths.

Credit: NOAA

reflect, and scatter different wavelengths of electromagnetic radiation, we can learn about the ocean by measuring the wavelengths of light emitted by its surface. Sensors that detect different types of surface-emitted radiation are known as **passive sensors**.

On April 1, 1960, NASA launched the world's first weather satellite, the TIROS-1 (Television Infrared Observation Satellite), one of a series of satellites carrying television cameras into space. Images of cloud and weather patterns ushered in an era of space-based weather forecasting. TIROS set the stage for the Earth-observing satellite systems that now deliver weather information every hour of every day across the globe.

Figure 4.48 NASA technicians prepare TIROS-1, the predecessor of TIROS-N and the world's first weather satellite, launched in 1960.

Credit: NASA

Scientists soon developed additional tools for space-based observations of Earth. Measurements of the temperature of the sea surface—a property known as **sea surface temperature** (SST)—began with the Nimbus satellites launched between 1964 and 1978. These satellites carried high-resolution infrared radiometers designed to measure the temperatures of cloud tops and the sea surface at night (when sunlight wouldn't interfere with the measurement). In 1978 NASA launched TIROS-N (N for NOAA), which carried the first of a series of Advanced Very-High-Resolution Radiometers (AVHRRs), multichannel, infrared-sensing instruments for measuring SST. The data from AVHRRs flown on TIROS-N and subsequent satellites represent the longest near-continuous

Figure 4.49 The Nimbus-1 satellite illustrated here is considered "the granddaddy of the current Earth-observing fleet" (Patel and Carlowicz 2023).

Credit: NASA

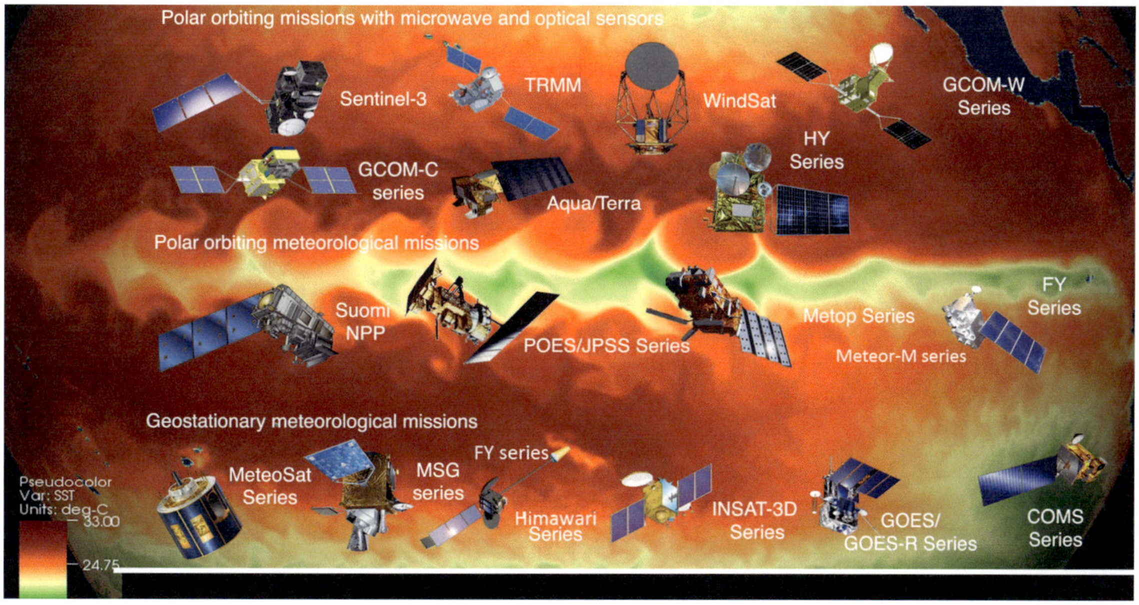

Figure 4.50 More than a dozen past and current satellite missions have been dedicated to measuring the temperature of the surface of the ocean.

Credit: O'Carroll CC BY 4.0.

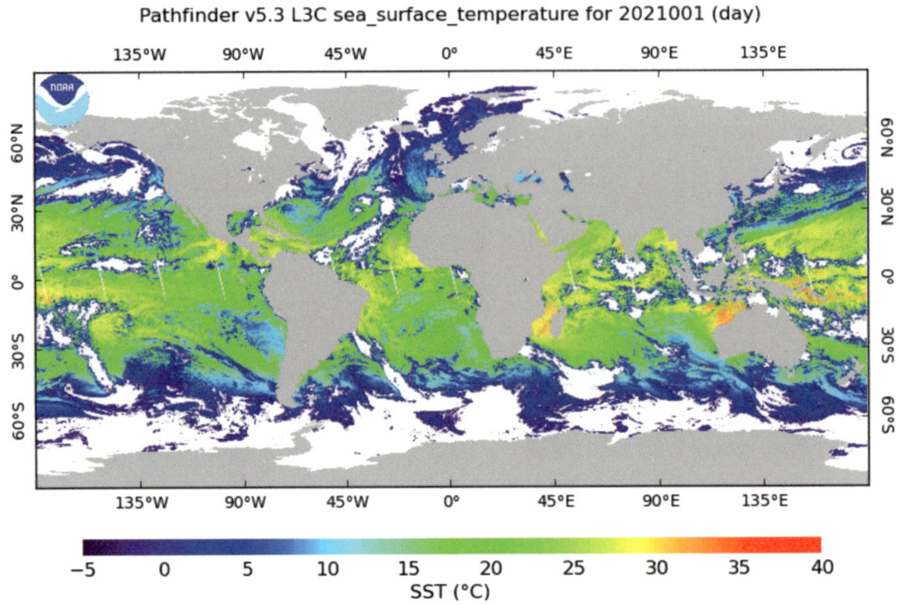

Figure 4.51 SST maps based on infrared measurements show gaps (white patches) where clouds occur. The color bar indicates the temperature (°C) of the different colors for interpreting the map.

***Credit*: National Centers for Environmental Information/NASA**

measurements of SST from space, a period of more than 40 years (e.g., Casey et al. 2010; O'Carroll et al. 2019; Minnett et al. 2019).

One limitation of AVHRR sensors is their inability to "see" through clouds, which block surface-emitted infrared radiation. To overcome this limitation, NASA and the Japanese Aerospace Exploration Agency launched satellites that carried **passive microwave sensors**. These sensors detect microwave radiation which passes through clouds and from which estimates of SST and other properties can be derived. Unfortunately, microwave sensors lack the resolution of infrared sensors and tend to be less accurate (Castro et al. 2008). So oceanographers take the best of both worlds and blend data from infrared and microwave sensors into a combined product, the Multi-scale Ultra-high-Resolution SSTs (MUR-SSTs). Global and regional maps of MUR-SSTs adhere to standards developed by the Group for High Resolution Sea Surface Temperature, which promotes best practices for monitoring, processing, and reporting satellite-derived SST (Minnett et al. 2019).

SST measurements serve operational oceanography, climate change research, fisheries, marine heatwave tracking, and other interests (e.g.,

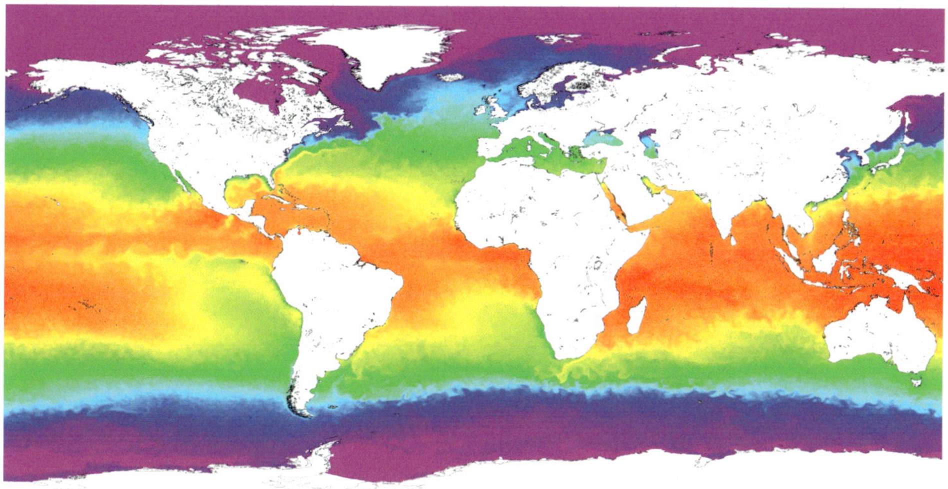

Figure 4.52 Satellite-derived SSTs are now routinely uploaded to the internet where they are available to the public.

Credit: NOAA/Office of Satellite and Product Ooperations

Beggs 2010; O'Carroll et al. 2019). NOAA's Coral Reef Watch uses satellite SST data to inform resource managers, scientists, and the public about the potential for **coral bleaching**, the phenomenon whereby coral animals expel their algal "helpers" in response to prolonged elevated seawater temperatures. In recent decades, coral bleaching has devastated corals around the world (e.g., Hughes et al. 2018). Satellite-based forecasts of coral bleaching allow scientists to "rescue" rare coral species and house them in a laboratory. When the warm waters subside, the corals can be returned to their natural environment. It's not the ideal solution, but coral rescues play an important role in coral conservation and management (e.g., Vardi et al. 2021).

Figure 4.53 Accurate and timely maps of SST are critical for protecting coral reefs like this one in Misool, Indonesia.

Credit: Ocean Image Bank/Alex Mustard

4.11.2 Sea Ice Extent

Soon after the *Titanic* sank following its collision with an iceberg in 1912, the governments of the US and Europe formed the International Ice Patrol. Tasked with monitoring the whereabouts of **icebergs**, seaborne chunks of fractured glaciers, the Ice Patrol was the first to regularly monitor ice conditions at polar latitudes. Interest in polar ocean ecosystems and climate change led to efforts in the 1960s to develop tools for observing icebergs and **sea ice**, the frozen surface of polar oceans (Zwally et al. 1983). Fortunately, passive microwave sensors proved ideal for the job.

Observations of sea ice using passive microwave sensors aboard aircraft began in the Arctic in 1967 (Gloersen et al. 1992). The first passive microwave instruments were flown on Russian satellites in 1968 (Tikhonov et al. 2016). But the Electrically Scanning Microwave Radiometer flown aboard the Nimbus 5 satellite in 1972 began the era of "all-weather, all-seasons" sea ice observations that has continued to this day (e.g., Parkinson et al. 1987). A 40-year record from passive microwave sensors has revealed striking and alarming retreats of sea ice in the Arctic and Antarctic Oceans (Parkinson 2019). Multiyear ice, sea ice that remains

Figure 4.54
The International Ice Patrol uses C-130 aircraft equipped with side-looking and forward-looking radar to detect, measure, and report icebergs in the North Atlantic.

Credit: US Coast Guard/Brandon Brewer

Figure 4.55 Passive microwave measurements of sea ice since the 1970s have proved invaluable for documenting the long-term decline of Arctic sea ice.

***Credit:* NASA Scientific Visualization Studio/Trent Schindler**

present for longer than one year—thought to be critical for maintaining sea ice volume—has shown similar declines (Haibo et al. 2020).

A desire for higher-resolution imagery led to the launch of **active microwave sensors**, instruments that produce their own electromagnetic signal and detect the reflected signal (like the radar gun used to catch speeders). By bouncing a beam of energy off the surface of the ocean, oceanographers can measure a number of ocean properties with very high resolution (e.g., Gens 2008). An active microwave system known as **synthetic aperture radar** (SAR) flew aboard Seasat and performed so well it inspired a generation of SAR-carrying satellites in the 1980s and 1990s (Tsatsoulis and Kwok 1998). SAR works by electronically mimicking a large antenna, enabling it to produce detailed images of sea ice and other properties. SAR remains an integral part of Earth-observing satellite systems in the 21st century.

Figure 4.56 Expanding melt ponds on the surface of the Arctic sea ice illustrate one of the consequences of a warmer planet.

***Credit:* NASA/ *Kathryn Hansen* CC BY 2.0**

4.11.3 Sea Surface Height

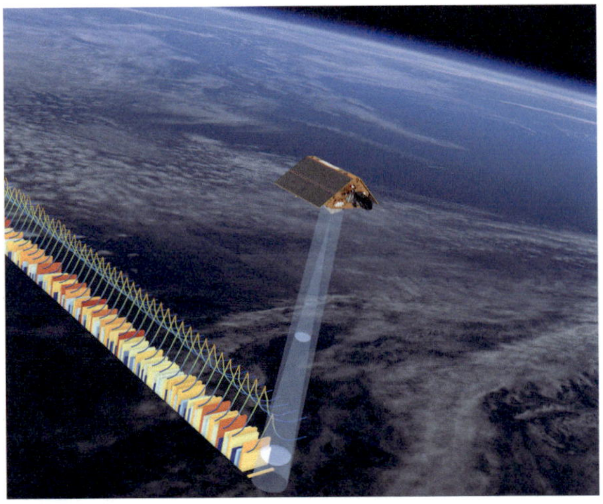

Figure 4.57 Satellite altimeters like the Sentinel-6 Michael Freilich measure the roundtrip time for a microwave pulse to generate maps of the bumps and dips of the sea surface.

Credit: European Space Agency

By bouncing a beam of energy off the surface of the ocean, oceanographers can measure the shape of the sea surface–the **sea surface topography**, the bumps and depressions of the ocean surface. The best-known devices, the **satellite altimeters**, measure the time it takes for a microwave radar signal to bounce off the ocean surface and return to the satellite. Based on the return time and precise knowledge of the height and path of the satellite, oceanographers can calculate the distance between the sea surface and the satellite.

Satellite altimeter measurements allow mapping of **sea surface height**, the height of the sea surface relative to a baseline, such as Earth's ellipsoid or geoid (i.e., its equi-gravitational surface; see Chapter 20).

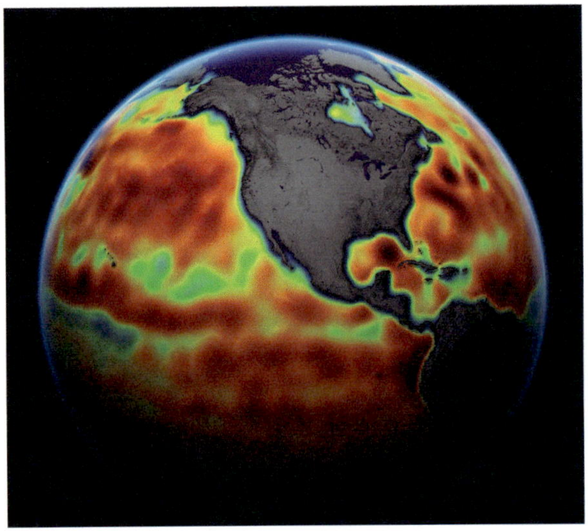

Figure 4.58 Sentinel-6 is already providing unprecedented views of the ocean's surface. Red colors indicate areas where sea level is higher than normal. Blue areas are lower than average.

Credit: NASA

Sea surface height measurements permit oceanographers to learn about ocean circulation and ocean tides. It also provides knowledge of sea level changes in response to a warming ocean and melting ice caps. As you might expect, these observations and the computations required to generate useful **satellite products**—the results provided by satellite measurements—involve very accurate measurements and lots of computing power.

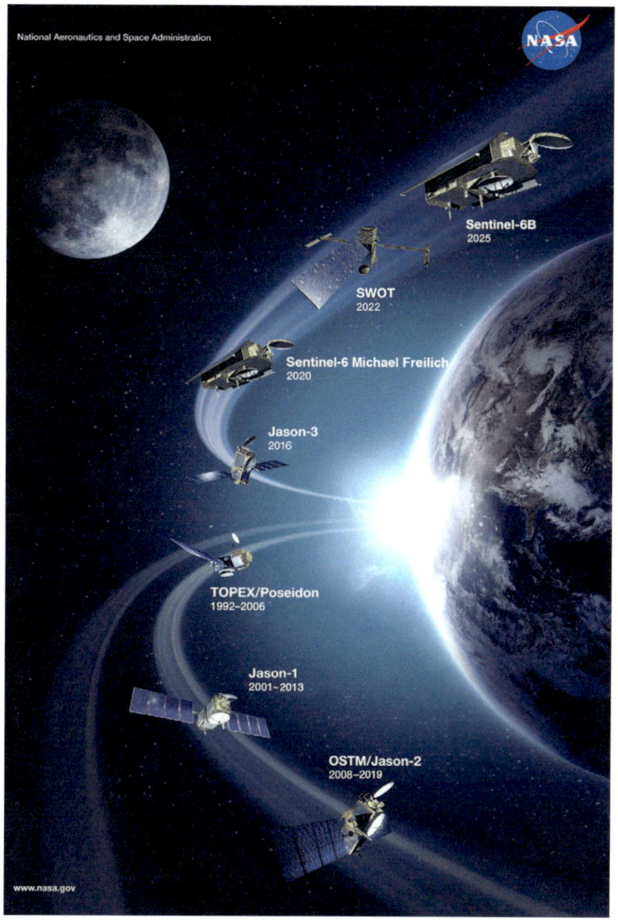

Credit: NASA

Figure 4.59 Satellites have measured sea surface height using altimeters since TOPEX/Poseidon in 1992.

Arguably the most successful oceanographic satellite altimeter mission to date is TOPEX/Poseidon, the Ocean Topography Experiment/Poseidon mission. Launched in 1992 in a collaboration between NASA and the French space agency, Centre National d'Etudes Spatiales, TOPEX/Poseidon allowed oceanographers to observe changes in world ocean circulation over seasonal and annual cycles, advanced our understanding of tides and sea level rise, and enabled significant refinement of general circulation models. It provided unprecedented observations of the 1997–1998 El Niño, among the strongest in 100 years. After more than 13 years and nearly 62,000 orbits of Earth, TOPEX/Poseidon ended operations in October 2005. Legendary oceanographer Walter Munk (1917–2019) called it "the most successful ocean experiment of all times" (Munk 2002).

The Jason satellite altimetry missions extended the data record begun by TOPEX/Poseidon. Jason-3 (launched in 2016) currently operates in tandem with the altimeter-carrying Sentinel-6 Michael Freilich satellite. This satellite was named after the former director of NASA's Earth Science Division, Michael Freilich (1954–2020)—a "passionate advocate" of measurements of Earth from space (Cook 2020). This satellite can track sea level rise with an error of less than a tenth of an inch (<1 mm) per year (Donlon et al. 2021).

4.11.4 Ocean Color

If you've ever spent time near an ocean in middle latitudes, you've undoubtedly noticed that the color of the water changes from season to season. Here in Southern California, the winter ocean appears blue, turns green in spring, blues up in summer, and greens again slightly in fall. These changes in ocean color originate from changes in the abundance of phytoplankton and their primary light-absorbing pigment, **chlorophyll**. Think green trees in summer and fallen leaves in winter. It's a similar phenomenon in the ocean. As the concentrations of phytoplankton vary, so does the color of the water: more phytoplankton, greener water. Viewed from space, these organisms give the ocean a palette of greenish hues in what might otherwise be a blue desert.

Figure 4.60
Green waters (top image) owe their hue to the absorption properties of phytoplankton, which largely absorb all colors except green. Blue waters (bottom image) owe their color to a lack of phytoplankton and the absorption properties of seawater, which absorbs mostly all colors except blue.

Credit: *Top: istock/richcarey; Bottom: istock/ShinOkamoto*

Oceanographers define **ocean color** as the "relative amounts of water-leaving radiance in the various portions of the visible spectrum" (Austin 1993). In other words, ocean color is the color of the ocean. Pure ocean water absorbs mostly every color except blue. That's why the open ocean appears blue. It's the only color remaining after all others are absorbed. The CZCS (mentioned above) provided biological oceanographers their first

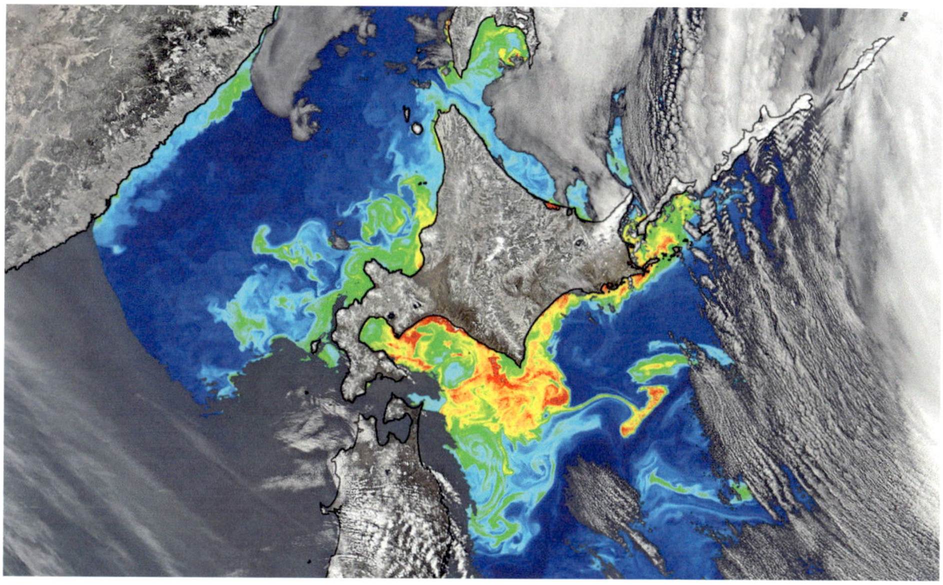

Figure 4.61 Swirls of red, orange, yellow, and green indicate high concentrations of phytoplankton in the currents around Hokkaido, Japan, on April 3, 2009.

Credit: NASA Goddard Space Flight Center

global views of ocean color, whose hues varied with the concentration of phytoplankton in the water. Images from the CZCS revealed an intimate connection between ocean physics (i.e., seasonal stratification, upwelling, eddies) and phytoplankton.

The CZCS greatly exceeded expectations and outlived its life expectancy, but it ceased operations in 1986. A decade later, the Sea-viewing Wide Field-of-view Sensor (SeaWiFS) lifted off from Vandenberg Air Force Base (August 1, 1997). By this time, oceanographers had developed in-water optical sensors and improved biochemical analyses of seawater. A new generation of oceanographers, the **bio-optical oceanographers**, stepped in. Their mission was to advance our understanding of biological oceanography through a study of ocean optics. Of them, Barale writes: "The space oceanographers . . . were like those pioneers who opened new trade routes, as what they discovered . . . surpassed many times what was expected in the beginning" (Barale 2010).

The color of the ocean reveals a surprising amount of information about its geology, chemistry, physics, and biology.

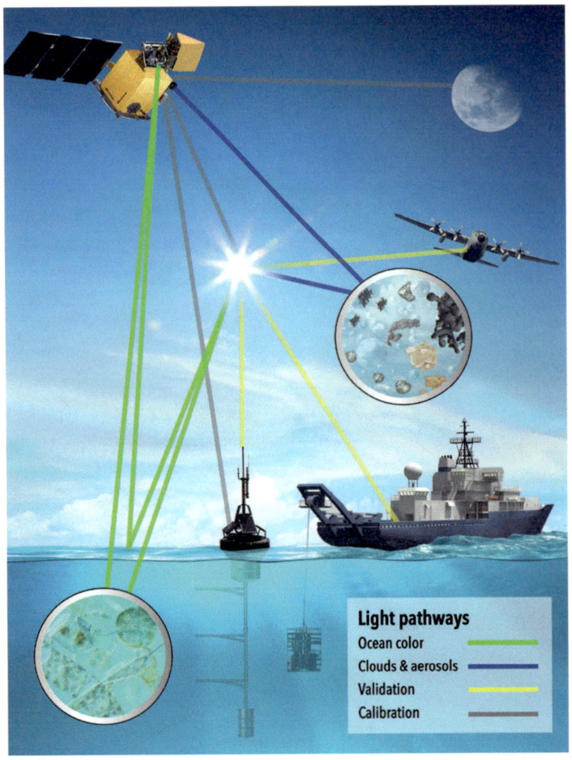

Figure 4.62 Satellites above and bio-optical instruments below establish the relationships between the optical properties of the ocean's biota and their changes over time.

Credit: NASA Goddard Space Flight Center

The design and execution of SeaWiFS built on the lessons learned from the CZCS, including the need for continuous **ground truthing**, the practice of verifying satellite observations with measurements obtained from within the water column. Though traditionally carried out on ships, ground truthing with AUVs is now under development (e.g., Bailey and Werdell 2006; Bennion et al. 2019). Buoys are also used, such as the Marine Optical Buoy system moored off the coast of Hawaii (e.g., Brown et al. 2007).

More than a dozen missions have been launched since SeaWiFS. The Moderate Resolution Imaging Spectroradiometer (MODIS) launched aboard the Terra satellite (December 1999) and the Aqua satellite (May 2002) brought improvements to sensor design and performance. The Visible and Infrared Imager/Radiometer Suite (VIIRS), the successor to MODIS, aboard the Suomi National Polar-orbiting Partnership, and a host of international sensors and satellites devoted to ocean color measurements ensure that global measurements of the ocean biosphere continue to this day.

Satellite measurements require regular ground truthing to verify that what they're seeing is actually what's being seen.

4.12 Ocean Observatories

If robots had underwater cities, they might look like **ocean observatories**, an integrated array of ocean-observing sensors and platforms on and above the seafloor designed to address questions of scientific and societal importance (e.g., Crise et al. 2018). Ocean observatories were conceived in the 1980s by American geological oceanographers as the oceanographer's version of an astronomical or volcano observatory (Delaney et al. 1987). Like these observatories, ocean observatories carry out observations over broad spatial and temporal scales. Uniquely, however, ocean observatories span hundreds of miles in waters thousands of feet deep. They involve fixed and mobile platforms connected either physically via cables or virtually via acoustic or optical signals. At heart, ocean observatories provide long-term and sustained observations of the ocean, in contrast to the brief encounters offered by shipboard oceanography.

To get an idea of what they look like, imagine an oceanic neighborhood with strings of instrumented moorings and AUV docking stations spaced at regular intervals like gas stations. Bottom-attached

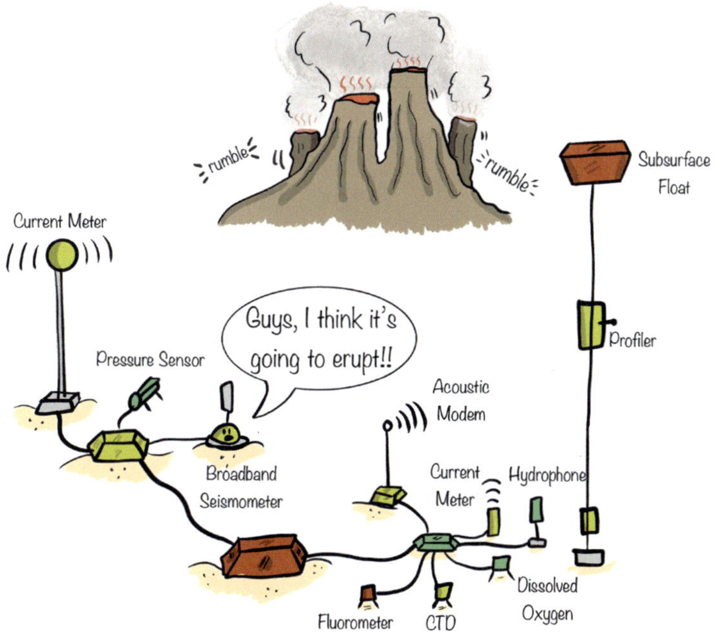

Figure 4.63 Artist's depiction of an ocean observatory.

Credit: Shaw and Chamberlin

Figure 4.64 A stack of thermometers measure the three-dimensional temperature structure of seawater as it flows out of a hot seafloor vent.

Credit: NSF-OOI/University of Washington/ Canadian Scientific Submersible Facility

instruments of all shapes and sizes cover the seafloor like tiny houses. Attached cables act as their power lines. Dish-shaped antennae transmit and receive acoustic and optical data. Drifters and gliders and AUVs fly through the surrounding water like a scene from *The Jetsons* (Hanna-Barbera 1962–1963). Bottom-roaming benthic explorers excavate sediments and retrieve and deploy instrument packages like All-Terrain Walkers on an undersea Tatooine (e.g., Lucas 1977–2019). Of course, the occasional human visitor interrupts the robotic seascape—arriving via ship or submersible or as a deep-sea diver—acting as a necessary overlord to keep the city operating smoothly.

Ocean observatories generally fall into one of three categories, regional cabled arrays, coastal arrays, and global arrays. Each serves different goals and purposes.

Figure 4.65 An octopus visits an observatory on the Axial Seamount off the coast of Oregon.

Credit: NSF-OOI/University of Washington/ Canadian Scientific Submersible Facility

Regional cabled arrays represent what geologists envisioned when they proposed ocean observatories in the 1980s. These arrays use optical-electrical telecommunication cables to provide

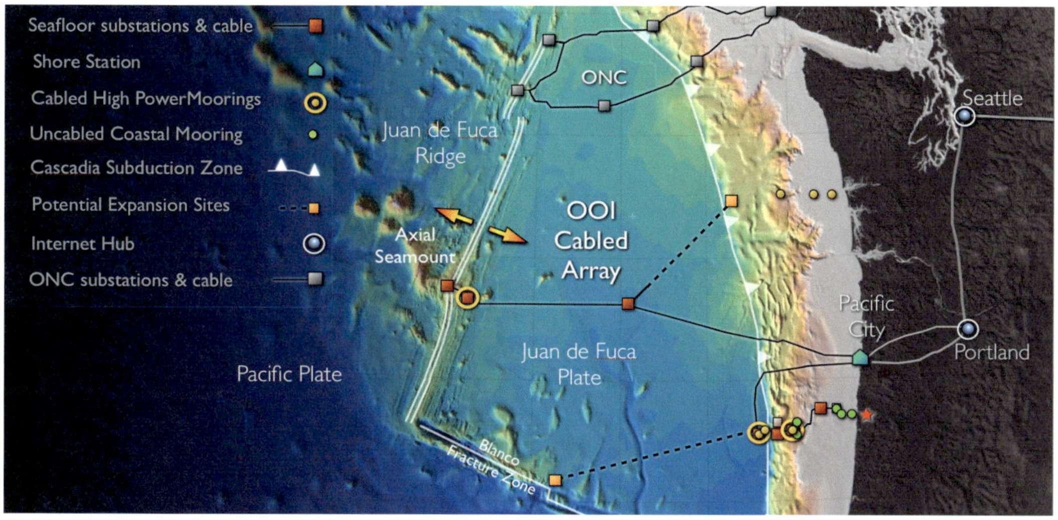

Figure 4.66 The regional cabled array along the Juan de Fuca Ridge includes instruments designed to study seafloor geology and to provide early warning of possible hazardous events, such as tsunami.

Credit: NSF-OOI/UW/CSSF

power and two-way communication capabilities to suites of instruments on the seafloor. The regional cabled array off the coasts of Washington and Oregon serves as the best example. This array, touted as "the first US ocean observatory to span a tectonic plate," consists of seven primary nodes, each with its own set of instruments and name (such as International District Vent Field 1). One array, the Axial Caldera Cabled Array, captured the eruption of an active undersea volcano on the Juan de Fuca Ridge on April 24, 2015 (Delaney et al. 2016).

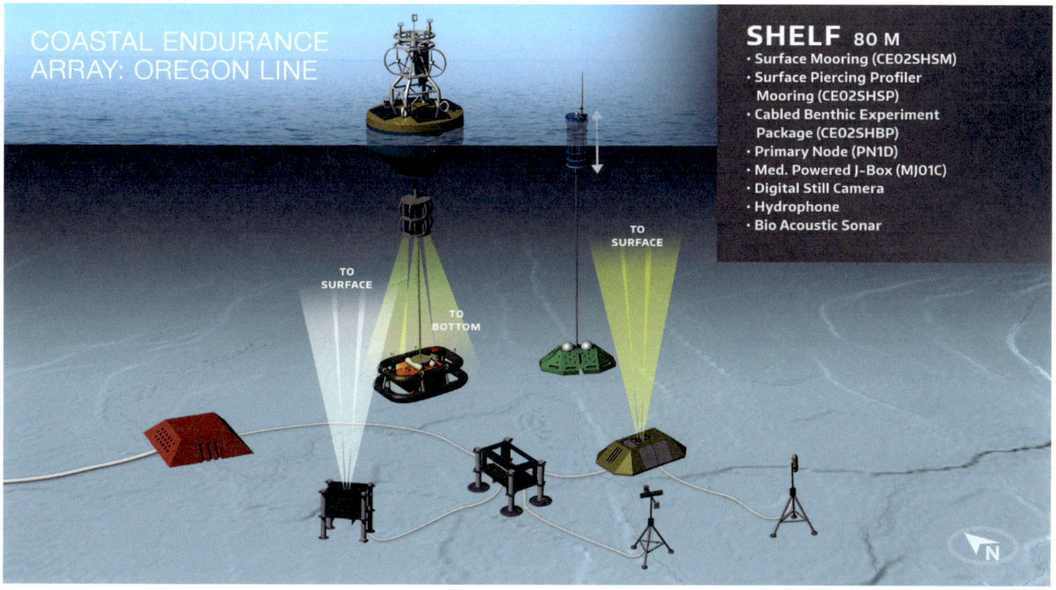

Figure 4.67 The Coastal Endurance Array along the Oregon continental shelf is the shallowest of the cabled observatories at 262 feet (80 m).

Credit: NSF-OOI/UW/CSSF

Coastal arrays focus on oceanographic processes that occur on or above the continental shelf (e.g., Trowbridge et al. 2019). A coastal array off the Oregon Shelf, the Coastal Endurance Array, employs moorings (fixed and profiler) in concert with gliders to provide high-resolution, top-to-bottom observations of ocean and atmospheric processes. The Coastal Endurance Array has observed hypoxia events (i.e., low oxygen), ocean heatwaves, and the response of marine organisms to a solar eclipse (Barth et al. 2018).

Global arrays refer to ocean observatories in select deep water locations that shed light on major science themes, such as air–sea exchange, climate variability, ocean circulation, and ocean ecosystems. Global arrays rely on deep ocean moorings and gliders to sample vertical and horizontal variability in the upper ocean. The Global Irminger Sea Array off Greenland has already provided tantalizing new observations that may change how we view circulation in this part of the world ocean (de Jong et al. 2018).

Ocean observatories represent an emerging conceptual, technological, and strategic shift in how oceanographers carry out

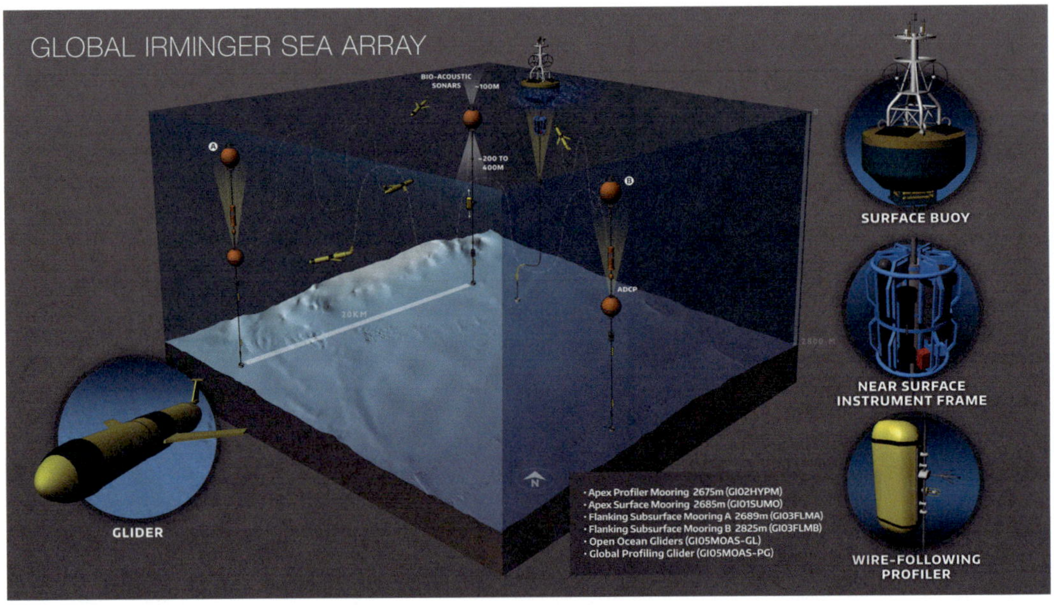

Figure 4.68
The Global Irminger Sea Array off the east coast of Greenland endures some of the harshest ocean conditions on the planet.

Credit: Image from work supported by the Ocean Observatories Initiative, a major facility fully funded by the National Science Foundation.

> **If we are to live on this planet responsibly and sustainably, then we must commit to building and maintaining global observing systems that take the pulse of our planet in real time.**

their work. But there are concerns that the enormous financial outlay needed to build and sustain ocean observatories will drain resources from traditional approaches (Witze 2013; Kintisch 2015). As stated by the National Research Council (2015), a "healthy balance" should be sought between "infrastructure to provide access to the ocean and advance the science . . . [and] funds for scientists and trainees to conduct research and provide value for the infrastructure investment." How oceanographers and funding agencies achieve this balance may determine the long-term vitality of future oceanographic research.

As a final note, I want to emphasize that ocean observatories will prove a massive undertaking. They represent, perhaps, the culmination of nearly 150 years of oceanographic research that began with the *Challenger* expedition. Heightened awareness of the rapid changes in Earth's climate and the growing knowledge of multiple stressors on the marine environment makes the mission of ocean observatories ever more urgent (e.g., Speich et al. 2019). With the UN Decade of Ocean Science for Sustainable Development upon us (2021–2030), the opportunities for ocean scientists look ever more promising. As Visbeck (2018) put it:

> The increased awareness of the importance of the ocean to the future of humanity gives grounds for cautious optimism and motivation for ambitious multilateral cooperation. The scientific community has been given a stage on which to shine during the Decade of Ocean Science for Sustainable Development. Let us come together, respect our disciplines and agendas, but also be ready to embark on an exciting and transformative journey to realize the ocean we need for the future we want.

4.13 Chapter References

Adams, Douglas. 1995. *The Hitchhiker's Guide to the Galaxy: A Trilogy in Five Parts*. London: Heinemann. https://www.penguin.co.uk/books/104/1041878/the-hitch-hiker-s-guide-to-the-galaxy/9780434023394.html

Austin, R. W. 1993. "Optical Remote Sensing of the Oceans: BC (Before CZCS) and AC (After CZCS)." In *Ocean Colour: Theory and Applications in a Decade of CZCS Experience*. Edited by Vittorio Barale and Peter M. Schlittenhardt. Dordecht: Kluwer Academic Publishers. https://doi.org/10.1007/978-94-011-1791-3

Bailey, Sean W., and P. Jeremy Werdell. 2006. "A Multi-sensor Approach for the On-Orbit Validation of Ocean Color Satellite Data Products." *Remote Sensing of the Environment* 102(1–2): 12–23. https://doi.org/10.1016/j.rse.2006.01.015

Ballard, Robert D., and Christopher Drew. 2021. *Into the Deep: A Memoir from the Man Who Found Titanic*. Washington, DC: National Geographic. https://www.simonandschuster.co.uk/books/Into-the-Deep/Robert-D-Ballard/9781426220999

Barale, Vittorio. 2010. "Preface." In *Oceanography from Space: Revisited*. Edited by V. Barale, J. Gower, and L Alberotanza, 249–271. Dordrecht: Springer. https://doi.org/10.1007/978-90-481-8681-5_15

Barber, Richard T., and Anna K. Hilting. 2000. "Achievements in Biological Oceanography." In *50 Years of Ocean Discovery: National Science Foundation 1950–2000*. Washington, DC: National Academies Press. https://doi.org/10.17226/9702

Barth, John A., Jonathan P. Fram, Edward P. Dever, Craig M. Risen, Chris E. Wingard, Robert W. Collier, and Thomas D. Kearney. 2018. "Warm Blobs, Low-Oxygen Events, and an Eclipse: The Ocean Observatories Initiative Endurance Array Captures Them All." *Oceanography* 31(1): 90–97. https://doi.org/10.5670/oceanog.2018.114

Bazilchuk, Nancy. 2022. "'Pamela' Makes Studying the Ocean Easy and Affordable." Ecomagazine. July 29, 2022. https://ecomagazine.com/news/industry/pamela-makes-studying-the-ocean-easy-and-affordable

Beggs, Helen M. 2010. "Use of TIR from Space in Operational Systems." In *Oceanography from Space: Revisited*. Edited by V. Barale, J. Gower, and L. Alberotanza, 249–271. Dordrecht: Springer. https://doi.org/10.1007/978-90-481-8681-5_15

Bennion, David H., David M. Warner, Peter C. Esselman, Brett Hobson, and Brian Kieft. 2019. "A Comparison of Chlorophyll *a* Values Obtained from an Autonomous Underwater Vehicle to Satellite-Based Measures for Lake Michigan." *Journal of Great Lakes Research* 45(4): 726–734. https://doi.org/10.1016/j.jglr.2019.04.003

Bowen, Andrew D., Dana Yoerger, Chris Taylor, Robert McCabe, Jonathan Howland, Daniel Gomez-Ibanez, James C. Kinsey, Matthew Heintz, Glenn McDonald, Donald B. Peters, John Bailey, Eleanor Bors, Tim Shank, Louis L. Whitcomb, Stephen C. Martin, Sarah E. Webster, Michael V. Jakuba, Barbara Fletcher, Chris Young, James Buescher, Patricia Fryer, and Samuel Hulme. 2009. "Field Trials of the Nereus Hybrid Underwater Robotic Vehicle in the Challenger Deep of the Mariana Trench." *Oceans* 2009: 1–10. https://doi.org/10.23919/OCEANS.2009.5422311

Braginsky, Boris, Alon Baruch, and Hugo Guterman. 2020. "Development of an Autonomous Surface Vehicle Capable of Tracking Autonomous Underwater Vehicles." *Ocean Engineering* 197: 106868. https://doi.org/10.1016/j.oceaneng.2019.106868

Brown, Steven W., Stephanie J. Flora, Michael E. Feinholz, Mark A. Yarbrough, Terrence Houlihan, Darryl Peters, Yong Sun Kim, James L. Mueller, B. Carol Johnson, and Dennis K. Clark. 2007. "The Marine Optical Buoy (MOBY) Radiometric Calibration and Uncertainty Budget for Ocean Color Satellite Sensor Vicarious Calibration." *Proceedings* SPIE 6714: *Sensors, Systems, and Next-Generation Satellites* XI: 67441M. https://doi.org/10.1117/12.737400

Bryson, Mitch, Matthew Johnson-Roberson, Richard J. Murphy, and Daniel Bongiorno. 2013. "Kite Aerial Photography for Low-Cost, Ultra-high Spatial Resolution Multi-Spectral Mapping of Intertidal Landscapes." *PLoS One* 8(9): e73550. https://doi.org/10.1371/journal.pone.0073550

Cael, B. B., Kelsey Bisson, Maureen Conte, Manon T. Duret, Christopher L. Follett, Stephanie A. Henson, Makio C. Honda, Morten H. Iversen, David M. Karl, Richard S. Lampitt, Colleen B. Mouw, Frank Muller-Karger, Corinne A. Pebody, Kenneth L. Smith Jr., and David Talmy. 2021. "Open Ocean Particle Flux Variability from Surface to Seafloor." *Geophysical Research Letters* 48(9): e2021GL092895. https://doi.org/10.1029/2021GL092895

Casella, Elisa, Jan Dreschel, Christian Winter, Markus Benninghoff, and Alessio Rovere. 2020. "Accuracy of Sand Beach Topography Surveying by Drones and Photogrammetry." *Geo-Marine Letters* 40: 255–268. https://doi.org/10.1007/s00367-020-00638-8

Casey, Kenneth S., Tess B. Brandon, Peter Cornillon, and Robert Evans. 2010. "The Past, Present, and Future of the AVHRR Pathfinder SST Program." In *Oceanography from Space*. Edited by V. Barale, J. Gower, and L Alberotanza, 273–287. Dordrecht: Springer. https://doi.org/10.1007/978-90-481-8681-5_16

Castro, Sandra L., Gary A. Wick, Darren L. Jackson, and William J. Emery. 2008. "Error Characterization of Infrared and Microwave Satellite Sea Surface Temperature Products for Merging and Analysis." *Journal of Geophysical Research* 113: C03010. https://doi.org/10.1029/2006JC003829

Christ, Robert D., and Robert L. Wernli. 2014. *The ROV Manual*. Waltham: Butterworth-Heinemann. https://www.sciencedirect.com/book/9780080982885/the-rov-manual

Colavito, Jason. 2014. *Jason and the Argonauts through the Ages*. Jefferson: McFarland & Company. https://mcfarlandbooks.com/product/jason-and-the-argonauts-through-the-ages/

Connor, J., B. Champion, and M. A. Joordens. 2021. "Current Algorithms, Communication Methods and Designs for Underwater Swarm Robotics: A Review." *IEEE Sensors Journal* 21(1): 153–169. https://doi.org/10.1109/JSEN.2020.3013265

Conte, Maureen H., and J. C. Weber. 2015. "Particle Flux in the Deep Sargasso Sea: The 35-Year Oceanic Flux Program Time Series." *Oceanography* 27(1): 142–147. https://doi.org/10.5670/oceanog.2014.17

Cook, Terri. 2020. "Obituary: Michael Freilich (1954–2020)." European Geosciences Union. August 7, 2020. https://www.egu.eu/news/678/obituary-michael-freilich-1954-2020/

Crease, James. 1962. "Velocity Measurements in the Deep Water of the Western North Atlantic: Summary." *Journal of Geophysical Research* 67(8): 3173–3176. https://doi.org/10.1029/JZ067i008p03173

Crise, Alessandro, Maurizio Ribera d'Alcalà, Patrizio Mariana, George Petihakis, Julie Robodart, Daniele Iudicone, Ralf Bachmayer, and Francesca Malfatti. 2018. "A Conceptual Framework for Developing the Next Generation of Marine OBservatories (MOBs) for Science and Society." *Frontiers in Marine Science* 5: 318. https://doi.org/10.3389/fmars.2018.00318

de Jong, M. Femke, Marilena Oltmanns, Johannes Kartensen, and Laura de Steur. 2018. "Deep Convection in the Irminger Sea Observed with a Dense Mooring Array." *Oceanography* 31(1); 50–59. https://doi.org/10.5670/oceanog.2018.109

Delaney, John R., Fred N. Speiss, Sean C. Solomon, Robert Hessler, Jill L. Karsten, John A. Baross, Robin T. Holcomb, Denis Norton, Russell E. McDuff, Fred Sayles, John Whitehead Jr., Dallas Abbott, and LeRoy Olsen. 1987. "Scientific Rationale for Establishing Long-Term Ocean Bottom Observatory/Laboratory Systems." In *Marine Minerals: Advances in Research and Resource Assessment*. Edited by P. G. Teleki, M. R. Dobson, J. R. Moore, and U. von Stackelberg. NATO ASI Series 194: 389–411. Dordrecht: D. Reidel Publishing. https://doi.org/10.1007/978-94-009-3803-8_27

Delaney, John R., Deborah S. Kelley, Aaron Marburg, Mark Stoermer, Hunter Hadaway, Kim Juniper, and Friedrich Knuth. 2016. "Axial Seamount—Wired and Restless: A Cabled Submarine Network Enables Real-Time Tracking of a Mid-Ocean Ridge Eruption and Live Video of an Active Hydrothermal System Juan de Fuca Ridge, NE Pacific." OCEANS 2016 MTS/IEEE *Monterey*, 2016: 1–8. https://doi.org/10.1109/OCEANS.2016.7761484

Dickey, Tommy D. 1991. "The Emergence of Concurrent High-Resolution Physical and Bio-optical Measurements in the Upper Ocean and Their Applications." *Reviews of Geophysics* 29(3): 383–413. https://doi.org/10.1029/91RG00578

Donlon, Craig J., Robert Cullen, Luisella Giulicchi, Pierrik Vuilleumier, C. Richard Francis, Mieke Kuschnerus, William Simpson, Abderrazak Bouridah, Mauro Caleno, Roberta Bertoni, Jesus Rancaño, Eric Pourier, Andrew Hyslop, James Mulcahy, Robert Knockaert, Christopher Hunter, Alan Webb, Marco Fornari, Parag Vaze, Shanon Brown, Joshua Willis, Shailen Desai, Jean-Damien Desjonqueres, Remko Scharroo, Cristain Martin-Puig, Eric Leuliette, Alejandro Egido, Walter H. F. Smith, Pascal Bonnefond, Sophie Le Gac, Nicolas Picot, and Gilles Tavernier. 2021. "The Copernicus Sentinel-6 Mission: Enhanced Continuity of Satellite Sea Level Measurements from Space." *Remote Sensing of the Environment* 258: 112395. https://doi.org/10.1016/j.rse.2021.112395

D'Ortenzio, Fabrizio, Vincent Taillandier, Hervé Claustre, Louis Marie Prieur, Edouard Leymarie, Alexander Mignot, Antoine Poteau, Christophe Penkerc'h, and Catherine Marie Schmechtig. 2020. "Biogeochemical Argo: The Test Case of the NAOS Mediterranean Array." *Frontiers in Marine Science* 7: 120. https://doi.org/10.3389/fmars.2020.00120.

Du, Zengfeng, Xiong Zhang, Chao Lian, Zhendong Luan, Shichuan Xi, Lianfu Li, Liang Mu, Jianxing Zhang, Wenzao Zhou, Xiufeng Chen, Zhijun Lu, Chuanbo Wang, Yu Chen, Jun Yan, and Xin Zhang. 2023. "The Development and Applications of a Controllable Lander for In-Situ, Long-Term Observation of Deep Sea Chemosynthetic Communities." *Deep Sea Research Part I: Oceanographic Research Papers* 193: 103960. https://www.sciencedirect.com/science/article/pii/S0967063722002734?via%3Dihub

Fiori, Lorenzo, Ashray Doshi, Emmanuelle Martinez, Mark B. Orams, and Barbara Bollard-Breen. 2017. "The Use of Unmanned Aerial Systems in Marine Mammal Research." *Remote Sensing* 9(6): 543. https://doi.org/10.3390/rs9060543

Flemming, B. W. 1976. "Side-Scan Sonar: A Practical Guide." *International Hydrographic Review* 53(1): 65–82. https://journals.lib.unb.ca/index.php/ihr/article/view/23734

Garwood, Jessica C., Andrew J. Lucas, Perry Naughton, Matthew H. Alford, Paul L. D. Roberts, Jules S. Jaffe, Laura deGelleke, and Peter J. S. Franks. 2020. "A Novel Cross-Shore Transport Mechanism Revealed by Subsurface, Robotic Larval Mimics: Internal Wave Deformation of the Background Velocity Field." *Limnology and Oceanography* 65(7): 1456–1470. https://doi.org/10.1002/lno.11400

Gens, Rudi. 2008. "Oceanographic Applications of SAR Remote Sensing." *GIScience & Remote Sensing* 45(3): 275–305. http://dx.doi.org/10.2747/1548-1603.45.3.275

Gentemann, C. L., Joel P. Scott, Piero L. F. Mazzini, Cassia Pianca, Santha Akella, Peter J. Minnett, Peter Cornillon, Baylor Fox-Kemper, Ivona Cetinić, T. Mike Chin, Jose Gomez-Valdes, Jorge Vazquez-Cuervo, Vardis Tsontos, Lisan Yu, Richard Jenkins, Sebastien De Halleux, Dave Peacock, and Nora Cohen. 2020. "Saildrone: Adaptively Sampling the Marine Environment." *Bulletin of the American Meteorological Society* 101(6): E744–E762. https://doi.org/10.1175/BAMS-D-19-0015.1

Gerringer, Mackenzie E., Thomas Linley, Alan J. Jamieson, Erica Goetze, and Jeffrey C. Drazen. 2017. "*Pseudoliparis swirei* sp. nov.: A Newly-Discovered Hadal Snailfish (Scorpaeniformes: Liparidae) from the Mariana Trench." *Zootaxa* 4358(1): 161–177. https://doi.org/10.11646/zootaxa.4358.1.7

Giddens, Jonatha, Alan Turchik, Whitney Goodell, Michelle Rodriguez, and Denley Delaney. 2021. "The National Geographic Society Deep-Sea Camera System: Low-Cost Remote Video Survey Instrument to Advance Biodiversity Observation in the Deep Ocean." *Frontiers in Marine Science* 7: 601411. https://doi.org/10.3389/fmars.2020.601411

Gillis, Justin. 2014. "In the Ocean, Clues to Change." *New York Times*. August 11, 2014. https://www.nytimes.com/2014/08/12/science/in-the-ocean-clues-to-change.html

Gloersen, Per, William J. Campbell, Donald J. Cavalieri, Josefino C. Comiso, Claire L. Parkinson, and H. Jay Zwally. 1992. *Arctic and Antarctic Sea Ice, 1987–1987: Satellite Passive-Microwave Observations and Analysis*. Washington, DC: Scientific and Technical Information Branch, National Aeronautics and Space Administration. https://ntrs.nasa.gov/citations/19940012966

Glud, Ronnie N., Peter Berg, Bo Thamdrup, Morten Larsen, Heather A. Stewart, Alan J. Jamieson, Anni Glud, Kazumasa Oguri, Hamed Sanei, Ashley A. Rowden, and Frank Wenzhöfer. 2021. "Hadal Trenches Are Dynamic Hotspots for Early Diagenesis in the Deep Sea." *Communications Earth & Environment* 2(21): 1–8. https://doi.org/10.1038/s43247-020-00087-2

Goni, Gustavo J., Robert E. Todd, Steven R. Jayne, George Halliwell, Scott Glenn, Jili Dong, Ruth Curry, Ricardo Domingues, Francis Bringas, Luca Centurioni, Steven F. DiMarco, Travis Miles, Julio Morell, Luis Pomales, Hyun-Sook Kim, Pelle E. Robbins, Glen G. Gawarkiewicz, John Wilkin, Joleen Heiderich, Becky Baltes, Joseph J. Cione, Greg Seroka, Kelly Knee, and Elizabeth R. Sanabia. 2017. "Autonomous and Lagrangian Ocean Observations for Atlantic Tropical Cyclone Studies and Forecasts." *Oceanography* 30(2): 92–103. https://doi.org/10.5670/oceanog.2017.227

Good, William S., Bradley Mallare, Zachary Payne, Jordan Wachs, Charles Wamsley, Joy Fasnacht, Rolfe Bode, and Sebastian Padilla. 2018. "Demonstration of Persistent, High Resolution Remote Sensing from an Advanced Stratollite Platform." In 2018 AIAA *Information Systems-AIAA Infotech@Aerospace*, 0193. https://doi.org/10.2514/6.2018-0193

Grare, Laurent, Nicholas Statom, Nick Pizzo, and Luc Lenain. 2021. "Instrumented Wave Gliders for Air-Sea Interaction and Upper Ocean Research." *Frontiers in Marine Science* 8: 664728. https://doi.org/10.3389/fmars.2021.664728

Gray, Patrick Clifton, Gregory D. Larsen, and David W. Johnson. 2022. "Drones Address an Observational Blind Spot for Biological Oceanography." *Frontiers in Ecology and the Environment* 20(7): 413–421. https://doi.org/10.1002/fee.2472

Guinness World Records. 2013. "Longest Journey by an Unmanned Autonomous Surface Vehicle." Accessed June 6, 2022. https://www.guinnessworldrecords.com/world-records/longest-journey-by-an-unmanned-autonomous-surface-vehicle

Haibo, Bi, Yu Liang, Yunhe Wang, Xi Liang, Zehua Zhang, Tingqin Du, Qinglong Yu, Jue Huang, Mei Kong, and Haijin Huang. 2020. "Arctic Multiyear Sea Ice Variability Observed from Satellites: A Review." *Journal of Oceanology and Limnology* 38: 962–984. https://doi.org/10.1007/s00343-020-0093-7

Hughes, Terry P., Kristen D. Anderson, Sean R. Connolly, Scott F. Heron, James T. Kerry, Janice M. Lough, Andrew H. Baird, Julia K. Baum, Michael L. Berumen, Tom C. Bridge, Danielle C. Claar, C. Mark Eakin, James P. Gilmour, Nicholas A. J. Graham, Hugo Harrison, Jean-Paul A. Hobbs, Andrew S. Hoey, Mia Hoogenboom, Ryan J. Lowe, Malcolm T. McCulloch, John M. Pandolfi, Morgan Pratchett, Verena Schoepf, Gergely Torda, and Shaun K. Wilson. 2018. "Spatial and Temporal Patterns of Mass Bleaching of Corals in the Anthropocene." *Science* 359: 80–83. https://doi.org/10.1126/science.aan8048

Jaffe, Jules S., Peter J. S. Franks, Paul L. D. Roberts, Diba Mirza, Curt Schurgers, Ryan Kastner, and Adrien Boch. 2017. "A Swarm of Autonomous Miniature Underwater Robot Drifters for Exploring Submesoscale Ocean Dynamics." *Nature Communications* 8: 14189. https://doi.org/10.1038/ncomms14189

JAMSTEC. 2017. "Deepest Fish Ever Recorded—Documented at Depths of 8,178 m in Mariana Trench." Japan Agency for Marine-Earth Science and Technology (JAMSTEC). August 24, 2017. https://www.jamstec.go.jp/e/about/press_release/20170824/.

Jayne, Steven R., and Neil M. Bogue. 2017. "Air-Deployable Profiling Floats." *Oceanography* 30(2): 29–31. https://doi.org/10.5670/oceanog.2017.214

Jayne, Steven R., Dean Roemmich, Nathalie Zilberman, Stephen C. Riser, Kenneth S. Johnson, Gregory C. Johnson, and Stephen R. Piotrowicz. 2017. "The Argo Program: Present and Future." *Oceanography* 30(2): 18–28. https://doi.org/10.5670/oceanog.2017.213

Katzschmann, Robert K., Joseph DelPreto, Robert McMurdy, and Daniela Rus. 2018. "Exploration of Underwater Life with an Acoustically Controlled Soft Robotic Fish." *ScienceRobotics* 3(16): eaar3449. https://doi.org/10.1126/scirobotics.aar3449

Kintisch, Eli. 2015. "A Moment of Truth Arrives for U.S. Ocean Science." *Science* 347(6221): 463. https://doi.org/10.1126/science.347.6221.463

Klemas, Victor. 2015. "Coastal and Environmental Remote Sensing from Unmanned Aerial Vehicles: An Overview." *Journal of Coastal Research* 315(5): 1260-1267. http://dx.doi.org/10.2112/JCOASTRES-D-15-00005.1

Knauss, John A., and Newell Garfield. 2017. *Introduction to Physical Oceanography*, 3rd ed. Long Grove: Waveland Press. https://www.waveland.com/browse.php?t=504

Kukulya, Amy L., Roger Stokey, Carl Fesiter, Edgar Mauricio Hoyos-Padilla, and Gregory Skomal. 2016. "Multi-vehicle Autonomous Tracking and Filming of White Sharks *Carcharodon carcharias*." 2016 *IEEE/OES Autonomous Underwater Vehicles* (AUV): 423–430. https://doi.org/10.1109/AUV.2016.7778707

Lee, Craig M., Theresa Paluszkiewicz, Daniel L. Rudnick, Melissa M. Omand, and Robert E. Todd. 2017. "Autonomous Instruments Significantly Expand Ocean Observing: An Introduction to the Special Issue on Autonomous and Lagrangian Platforms and Sensors (ALPS)." *Oceanography* 30(2): 15–17. https://doi.org/10.5670/oceanog.2017.211

Li, Guorui, Xiangping Chen, Fanghao Zhou, Yiming Liang, Youhua Xiao, Xunuo Cao, Zhen Zhang, Mingqi Zhang, Baosheng Wu, Shunyu Yin, Yi Xu, Hongbo Dan, Zheng Chen, Wei Song, Wenjing Yang, Binbin Pan, Jiaoyi Hou, Weifend Zou, Shunping He, Xuxu Yang, Guoyong Mao, Zheng Jia, Haofei Zhou, Tiefeng Li, Shaoxing Qu, Zhongpin Xu, Zhilong Huang, Yingwu Luo, Tao Xie, Jason Gu, Shiqiang Zhu, and Wei Yang. 2021. "Self-Powered Soft Robot in the Mariana Trench." *Nature* 591: 66–71. https://doi.org/10.1038/s41586-020-03153-z

Li, Xue, Shaoling Shang, Zhongping Lee, Gong Lin, Yongnian Zhang, Jingyu Wu, Zhenjun Kang, Xiangxu Liu, Cheng Yin, and Yue Gao. 2021. "Detection and Biomass Estimation of *Phaeocystis globosa* Blooms off Southern China from UAV-Based Hyperspectral Measurements." *IEEE Transactions on Geoscience and Remote Sensing* 60: 1–13. https://doi.org/10.1109/TGRS.2021.3051466

Lindsey, Rebecca, and Luann Dahlman. 2020. "Climate Change: Ocean Heat Content." NOAA Climate. August 17, 2020. https://www.climate.gov/news-features/understanding-climate/climate-change-ocean-heat-content

Lipsett, Lonny. 2014. "A Long-Lost Legacy of Ocean Research Resurfaces." *Oceanus*. February 6, 2014. https://www.whoi.edu/oceanus/feature/bumpus/

Lumberg, Johannes, Frank Wenzhöfer, Machael Hofbauer, Paul Färber, and Volker Meyer. 2018. "Benthic Crawler NOMAD." 2018 OCEANS - *MTS/IEEE Kobe Techno-Oceans* (OTO), 1–7. https://doi.org/10.1109/OCEANSKOBE.2018.8559446

Lumpkin, Rick, Tamay Özgökmen, and Luca Centurioni. 2017. "Advances in the Application of Surface Drifters." *Annual Review of Marine Science* 9: 59–81. https://doi.org/10.1146/annurev-marine-010816-060641

Marx, Robert F. 1990. *The History of Underwater Exploration*. Kiribati: Dover Publications. https://www.google.com/books/edition/The_History_of_Underwater_Exploration/oiWFhoRzPBQC?hl=en

Mazzolai, Barbara, Federico Carpi, Koichi Suzumori, Matteo Cianchetti, Thomas Speck, Stoyan K. Smoukov, Ingo Burgert, Tobias Keplinger, Gilberto Siqueira, Felix Vanneste, Oliver Goury, Christian Duriez, Thrishantha Nanayakkara, Bram Vanderborght, Joost Brancart, Seppe Terryn, Steven Rich, Ruituan Liu, Kenjiro Fukuda, Takao Someya, Marcello Calisti, Cecilia Laschi, Wenguang Sun, Gang Wang, Li Wen, Robert Baines, Sree Kalyan Patiballa, Rebecca Kramer, Daniela Rus, Peer Fischer, Friedrich Simmel, and Andreas Lendlein. 2022. "Roadmap on Soft Robotics: Multifunctionality, Adaptability, and Growth Without Borders." *Multifunctional Materials* 5(3): 032001. https://doi.org/10.1088/2399-7532/ac4c95

Meyer, H. K., E. M. Roberts, H. T. Rapp, and A. J. Davies. 2019. "Spatial Patterns of Arctic Sponge Ground Fauna and Demersal Fish Are Detectable in Autonomous Underwater Vehicle (AUV) Imagery." *Deep Sea Research Part I: Oceanographic Research Papers* 153: 103137. https://doi.org/10.1016/j.dsr.2019.103137

Minnett, P. J., A. Alvera-Azcárate, T. M. Chin, G. K. Corlett, C. L. Gentemann, I. Karagali, X. Li, A. Marsouin, S. Marullo, E. Maturi, R. Santoleri, S. Saux Picart, M. Steele, and J. Vazquez-Cuervo. 2019. "Half a Century of Satellite Remote Sensing of Sea-Surface Temperature." *Remote Sensing of the Environment* 233: 111366. https://doi.org/10.1016/j.rse.2019.111366

Molar-Candanosa, Robert. "NASA Is Helping Fly Drones in the Arctic. Here's What That Means for Sea Ice and Sea Level Change." NASA. February 24, 2022. https://www.nasa.gov/feature/goddard/2022/nasa-helps-fly-drone-in-alaska-to-study-sea-ice-thickness

Mongilio, Heather. 2023. "Titan Submersible Debris Found Near Titanic Wreck; Navy Sensors Detected Implosion." USNI News. June 22, 2023. https://news.usni.org/2023/06/22/coast-guard-titan-submersible-suffered-catastrophic-loss-debris-found-near-titanic-wreck

Moore, Steven W., Harry Bohm, and Vickie Jensen. 2010. *Underwater Robotics: Science, Design & Fabrication*. Monterey: Marine Advanced Technology Education Center. https://www.marinetech.org/underwater-robotics-textbook-2/

Munk, Walter. 2002. Testimony to the US Commission on Ocean Policy. San Pedro, California. April, 18, 2002. https://govinfo.library.unt.edu/oceancommission/meetings/apr18_19_02/apr18_19_02.html#agenda

Murfitt, Sarah L., Blake M. Allan, Alecia Bellegrove, Alex Rattray, Mary A. Young, and Daniel Ierodianconou. 2017. "Applications of Unmanned Aerial Vehicles in Intertidal Reef Monitoring." *Scientific Reports* 7: 10259. https://doi.org/10.1038/s41598-017-10818-9

Nakajoh, Hidehiko, Takashi Murashima, and Fumitaka Sugimoto. 2018. "Development of Full Depth Fiber Optic Cable ROV (UROV11K) System." OCEANS 2018 MTS/IEEE *Charleston*: 1–8. https://doi.org/10.1109/OCEANS.2018.8604795

National Research Council. 2015. *Sea Change: 2015-2025 Decadal Survey of Ocean Sciences*. Washington, DC: National Academies Press. https://doi.org/10.17226/21655

Nehring, Abigail. 2021. "Douglas Webb: Tinkerer, Engineer and Oceanographer Emeritus." The Enterprise. January 26, 2021. https://www.capenews.net/falmouth/news/douglas-webb-tinkerer-engineer-and-oceanographer-emeritus/article_c8fd4ab2-bdb0-57e5-b0df-373f47fb14d5.html

Newpower, Anthony. 2006. *Iron Men and Tin Fish: The Race to Build a Better Torpedo During World War II*. Westport: Praeger Security International. https://www.abc-clio.com/products/d1212c/

Nguyen, Kien, Clinton Fookes, Sridha Sridharan, Yingli Tian, Feng Liu, Xiaoming Liu, and Arun Ross. 2022. "The State of Aerial Surveillance: A Survey." *IEEE Transactions* arXiv:2201.03080v2. https://doi.org/10.48550/arXiv.2201.03080

Niiler, Peter. 2001. "The World Ocean Surface Circulation." In *Ocean Circulation and Climate: Observing and Modelling the Global Ocean*. Volume 77. Edited by Gerold Siedler, John Church, and John Gould, 193–204. Oxford: Academic Press. https://doi.org/10.1016/S0074-6142(01)80119-4

O'Carroll, Anne G., Edward M. Armstrong, Helen M. Beggs, Marouan Bouali, Kenneth S. Casey, Gary K. Corlett, Prasanjit Dash, Craig J. Donlon, Chelle L. Gentemann, Jacob L. Høyer, Alexander Ignatov, Kamila Kabobah, Misako Kachi, Yukio Kurihara, Ioanna Karagali, Eileen Maturi, Christopher J. Merchant, Salvatore Marullo, Peter J. Minnett, Matthew Pennybacker, Balaji Ramakrishnan, RAAJ Ramsankaran, Rosalia Santoleri, Swathy Sunder, Stéphane Saux Picart, Jorge Vázquez-Cuervo, and Irernfrid Wimmer. 2019. "Observational Needs of Sea Surface Temperature." *Frontiers of Marine Science* 6: 420. https://doi.org/10.3389/fmars.2019.00420

Pagniello, Camille M. L. S., Megan A. Cimino, and Eric Terrill. 2019. "Mapping Fish Chorus Distributions in Southern California Using an Autonomous Wave Glider." *Frontiers in Marine Science* 6: 526. https://doi.org/10.3389/fmars.2019.00526

Parkinson, Claire. 2019. "A 40-y Record Reveals Gradual Antarctic Sea Ice Increases Followed by Decreases at Rates Far Exceeding the Rates Seen in the Arctic." PNAS 116(29): 14414–14423. https://doi.org/10.1073/pnas.1906556116

Parkinson, Claire L., Josefino C. Comiso, H. Jay Zwally, Donald J. Cavalieri, and Per Gloersen. 1987. *Arctic Sea Ice, 1973–1976: Satellite Passive-Microwave Observations*. Washington DC: Scientific and Technical Information Branch, National Aeronautics and Space Administration. https://ntrs.nasa.gov/citations/19870015437

Patel, Kasha, and Mike Carlowicz. 2023. "The Legacy of Nimbus." NASA. Accessed October 4, 2023. https://earthobservatory.nasa.gov/images/84542/the-legacy-of-nimbus

Roemmich, Dean, Matthew H. Alford, Hervé Claustre, Kenneth Johnson, Brian King, James Moum, Peter Oke, W. Brechner Owens, Sylvie Pouliquen, Sarah Purkey, Megan Scanderberg, Toshio Suga, Susan Wijffels, Nathalie Zilberman, Dorothee Bakker, Molly Baringer, Mathieu Belbeoch, Henry C. Bittig, Emmanuel Boss, Paulo Calil, Fiona Carse, Thierry Carval, Fei Chai, Diarmuid Ó. Conchubhair, Fabrizio d'Ortenzio, Giorgio Dall'Olmo, Damien Desbruyeres, Katja Fennel, Ilker Fer, Raffaele Ferrari, Gael Forget, Howard Freeland, Tetsuichi Fujiki, Marion Gehlen, Blair Greeenan, Robert Hallberg, Toshiyuki Hibiya, Shigeki Hosoda, Steven Jayne, Markus Jochum, Gregory C. Johnson, KiRyong Kang, Nicolas Kolodziejczyk, Arne Körtzinger, Pierre-Yves Le Traon, Yueng-Djern Lenn, Guillaume Maza, Kjell Arne Mork, Tamaryn Morris, Takeyoshi Nagai, Jonathan Nash, Alberto Naveira Garabato, Are Olsen, Rama Rao Pattabhi, Satya Prakash, Stephen Riser, Catherine Schmechtig, Claudia Schmid, Emily Shroyer, Andreas Sterl, Philip Sutton, Lynne Talley, Toste Tanhua, Virginie Thierry, Sandy Thomalla, John Toole, Ariel Troisi, Thomas W. Trull, Jon Turton, Pedro Joaquin Velez-Belchi, Waldemar Walczowski, Haili Wang, Rik Wanninkhof, Amy F. Waterhouse, Stephanie Waterman, Andrew Watson, Cara Wilson, Annie P. S. Wong, Jianping Xu, and Ichiro Yasuda. 2019. "On the Future of Argo: A Global, Full-Depth, Multi-Disciplinary Array." *Frontiers in Marine Science* 6: 439. https://doi.org/10.3389/fmars.2019.00439

Sanabia, Elizabeth, and Steven R. Jayne. 2020. "Ocean Observations Under Two Major Hurricanes: Evolution of the Response Across the Storm Wakes." *AGU Advances* 1(3): e2019AV000161. https://doi.org/10.1029/2019AV000161

Schlee, John S., Herman A. Karl, and M.E. Torresan. 1995. *Imaging the Sea Floor*. (No. 2079). US Government Printing Office. https://doi.org/10.3133/b2079

Schofield, Oscar, Josh Kohut, David Aragon, Liz Creed, Josh Graver, Chip Haldeman, John Kerfoot, Hugh Roarty, Clayton Jones, Doug Webb, and Scott Glenn. 2007. "Slocum Gliders: Robust and Ready." *Journal of Field Robotics* 24(6): 473–485. https://doi.org/10.1002/rob.20200

Searle, R. C., C. J. MacLeod, C. Peirce, and T. J. Reston. 2018. "The Mid-Atlantic Ridge Near 13°20′N: High-Resolution Magnetic and Bathymetry Imaging." *Geochemistry, Geophysics, Geosystems* 20(1): 295–313. https://doi.org/10.1029/2018GC007940

Sivčev, Staja, Joseph Coleman, Edin Omerdić, Gerard Dooly, and Daniel Toal. 2018. "Underwater Manipulators: A Review." *Ocean Engineering* 163: 431–450. https://doi.org/10.1016/j.oceaneng.2018.06.018

Smith, Raymond C., and Karen S. Baker. 1978. Optical Classification of Natural Waters. *Limnology and Oceanography* 23(2): 260–267. https://doi.org/10.4319/lo.1978.23.2.0260

Speich, Sabrina, Tong Lee, Frank Muller-Karger, Laura Lorenzoni, Ananda Pascual, Di Jin, Eric Delory, Gilles Reverdin, John Siddorn, Marlon R. Lewis, Nuria Marba, Pier Luigi Buttigieg, Sanae Chiba, Justin Manley, Amos Tiereyangn Kabo-Bah, Kruti Desai, and Abby Ackerman. 2019. "Editorial: Oceanobs'19: An Ocean of Opportunity." *Frontiers in Marine Science* 6: 570. https://doi.org/10.3389/fmars.2019.00570

Stommel, Henry. 1965. *The Gulf Stream*. Berkeley: University of California Press. Reissued 2022. https://www.ucpress.edu/book/9780520318557/the-gulf-stream

Storch, Hans Von, Amders Omstedt, and Ulrich Callies. 2020. "Drift Bottles in the Southern Baltic Sea—an Experiment-of-Opportunity." Unpublished manuscript. April 2020. https://www.researchgate.net/publication/340885296_Drift_bottles_in_the_southern_Baltic_Sea_-_an_experiment-of-opportunity

Stuart, Mary B., Andrew J. S. McGonigle, and Jon R. Willmott. 2019. "Hyperspectral Imaging in Environmental Monitoring: A Review of Recent Developments and Technological Advances in Compact Field Deployable Systems." *Sensors* 19(14): 3071. https://doi.org/10.3390/s19143071

Swallow, John. 1955. "A Neutral Buoyancy Float for Measuring Deep Currents." *Deep Sea Research* 3: 74–81. https://doi.org/10.1016/0146-6313(55)90037-X

Tikhonov, V. V., M. D. Raev, E. A. Sharkov, D. A. Boyarskii, I. A. Repina, and N. Yu Komarova. 2016. "Satellite Microwave Radiometry of Sea Ice of Polar Regions: A Review." *Izvestiya, Atmospheric and Oceanic Physics* 52: 1012–1030. https://doi.org/10.1134/S0001433816090267

Trowbridge, John, Robert Weller, Deborah Kelley, Edward Dever, Albert Plueddemann, John A. Barth, and Orest Kawka. 2019. "The Ocean Observatories Initiative." *Frontiers of Marine Science* 6: 74. https://doi.org/10.3389/fmars.2019.00074

Tsatsoulis, Costas, and Ronald Kwok. 1998. "Recent Advances in the Analysis of SAR Data of the Polar Oceans." In *Analysis of SAR Data of the Polar Oceans*. Edited by Costas Tsatsoulis and Ronald Kwok. Heidelberg: Springer Berlin. https://doi.org/10.1007/978-3-642-60282-5

Vardi, Tali, Whitney C. Hoot, Jessica Levy, Elizabeth Shaver, R. Scott Winters, Anastazia T. Banaszak, Iliana B. Baums, Valéries F. Chamberland, Nathan Cook, David Gulko, Margaux Y. Hein, Les Kaufman, Michelle Loewe, Petra Lundgren, Caitlan Lustic, Petra MacGowan, Mikhail V. Matz, Miles McGonigle, Ian McLeod, Jennifer Moore, Tom Moore, Sandrine Pivard, F. Joseph Pollock, Baruch Rinkevich, David, J. Suggett, Samuel Suleiman, T. Shay Viehman, Tatiana Villalobos, Virginia M. Weis, Chelsea Wolke, and Phanor H. Montoya-Maya. 2021. "Six Priorities to Advance the Science and Practice of Coral Reef Restoration Worldwide." *Restoration Ecology* 29(8): e13498. https://doi.org/10.1111/rec.13498

Visbeck, Martin. 2018. "Ocean Science Research Is Key for a Sustainable Future." *Nature Communications* 9: 690. https://doi.org/10.1038/s41467-018-03158-3

Voosen, Paul. 2018. "Saildrone Fleet Could Help Replace Aging Buoys." *Science* 359(6380): 1082–1083. https://doi.org/10.1126/science.359.6380.1082

Wang, Jian, Yuan-gui Tang, Chuan-xu Chen, Ji-xu Li, Cong Chen, Ai-qun Zhang, Yi-ping Li, and Shuo Li. 2020. "Terrain Matching Localization for Hybrid Underwater Vehicle in the Challenger Deep of the Mariana Trench." *Frontiers of Information Technology & Electronic Engineering* 21: 749–759. https://doi.org/10.1631/FITEE.1900556

Witze, Alexandra. 2013. "Marine Science: Oceanography's Billion-Dollar Baby." *Nature* 501: 480–482. https://doi.org/10.1038/501480a

Woods Hole Oceanographic Institution. "Dean F. Bumpus." Accessed June 5, 2022. https://www.whoi.edu/who-we-are/about-us/people/obituary/dean-f-bumpus/

World View. 2023. "Engineering Peak Experiences in the Stratosphere." Accessed March 13, 2023. https://worldview.space/space-tourism/

Wu, Ziyin, Fanlin Yang, and Yong Tang. 2021. "Side-scan Sonar and Sub-bottom Profiler Surveying." In *High-resolution Seafloor Survey and Applications*, 95–122. Singapore: Springer. https://doi.org/10.1007/978-981-15-9750-3_4

Xiang, Xianbo, Zemin Niu, Lionel Lapierre, and Mingjiu Zuo. 2015. "Hybrid Underwater Robotic Vehicles: The State-of-the-Art and Future Trends." *HKIE Transactions* 22(2): 103–116. https://doi.org/10.1080/1023697X.2015.1038322

Yoerger, Dana R., Annette F. Govindarajan, Jonathan C. Howland, Joel K. Llopiz, Peter H. Wiebe, Molly Curran, Justin Fujii, Daniel Gomez-Ibanez, Kakani Katija, Bruce H. Robison, Brett W. Hobson, Michael Risi, and Stephen M. Rock. 2021. "A Hybrid Underwater Robot for Multidisciplinary Investigation of the Ocean Twilight Zone." *ScienceRobotics* 6(55): 1–12. https://doi.org/10.1126/scirobotics.abe1901

Zhang, Yanwu, John P. Ryan, Brett W. Hobson, Brian Kieft, Anna Romano, Benedetto Barone, Christina M. Preston, Brent Roman, Ben-Yair Raanan, Douglas Pargett, Mathilde Dugenne, Angelicque E. White, Fernanda Henderikx Freitas, Steve Poulos, Samuel T. Wilson, Edward F. DeLong, David M. Karl, James M. Birch, James G. Bellingham, and Christopher A. Scholin. 2021. "A System of Coordinated Autonomous Robots for Lagrangian Studies of Microbes in the Oceanic Deep Chlorophyll Maximum." *ScienceRobotics* 6(50): eabb9138. https://doi.org/10.1126/scirobotics.abb9138

Zhang, Yanwu, John P. Ryan, Brian Kieft, Brett W. Hobson, Robert S. McEwen, Michael A. Godin, Julio B. Harvey, Benedetto Barone, James G. Bellingham, James M. Birch, Christopher A. Scholin, and Francisco P. Chavez. 2019. "Targeted Sampling by Autonomous Underwater Vehicles." *Frontiers in Marine Science* 6: 415. https://doi.org/10.3389/fmars.2019.00415

Zinke, Julika, Matthew E. Salter, Caroline Leck, Michael J. Lawler, Grace C. E. Porter, Michael P. Adams, Ian M. Brooks, Benjamin J. Murray, and Paul Zieger. 2021. "The Development of a Miniaturised Balloon-Borne Cloud Water Sampler and Its First Deployment in the High Arctic." *Tellus B: Chemical and Physical Meteorology* 73(1): 1–12. https://doi.org/10.1080/16000889.2021.1915614

Zwally, H. Jay, Josefino C. Comiso, Claire L. Parkinson, William J. Campbell, Frank D. Carsey, and Per Gloersen. 1983. *Antarctic Sea Ice, 1973-1976: Satellite Passive-Microwave Observations*. Washington, DC: Scientific and Technical Information Branch, National Aeronautics and Space Administration. https://ntrs.nasa.gov/citations/19840002650

Credit: Shaw and Chamberlin

Figure 4.69
Perhaps one day ocean robots will even look and talk like us.

I hear fish talking.
Artwork by Martha Rich.

Acoustical Oceanography

5

FIGURE 5.1
Humpback whales sing like no other animal in the world. Their haunting melodies fill the ocean with sounds scientists have yet to decipher.

Credit: istock/Yann-HUBERT

If you've ever seen the movie *Star Trek IV: The Voyage Home* (Nimoy 1986), you undoubtedly would recognize the deep bass rumblings and high-pitched wailings of humpback whales. The sounds were recorded in the 1950s and 1960s by navy engineer Frank Watlington (1916–1982), who worked at a top-secret navy listening post on Bermuda. Watlington's job was to listen for Soviet submarines using an underwater microphone, or **hydrophone**. Hydrophones work just like an abovewater microphone except they're waterproof. But instead of submarines, he heard the eerie sounds of humpback whales. Word of the sounds reached American biologist Roger Payne (b. 1935), who recognized repetitive patterns in the vocalizations, which he called **songs** (Payne and McKay 1971). In 1970, Payne released these recordings on an LP, *Songs of the Humpback Whales*. It became the best-selling nature album of all time (Rothenberg 2014).

Whale songs are not the only sounds you will hear in the ocean. Underwater sounds come from a surprising variety of sources: bubbles, calving icebergs, rainfall, waves, underwater volcanoes, shrimp, crabs, fish, whales, dolphins, and ships, among many others. This is no "silent world," as suggested by the title of Jacques Cousteau's 1956 underwater

FIGURE 5.2
A hydrophone among the fishes at Grays Reef National Marine Sanctuary off the coast of Georgia.

Credit:* NOAA/*Jon Martinez

documentary, *Le Monde du Silence*. In fact, the ocean is quite noisy, rivaling the soundscapes you might hear in a tropical rainforest or a southern bayou.

Two subdisciplines of underwater acoustics have emerged in modern times: **active acoustics**, the use of artificially produced sounds to study ocean properties and detect underwater objects; and **passive acoustics**, the study of underwater sounds using listening devices (i.e., hydrophones). Oceanographers in the twenty-first century employ tools and techniques in both types of underwater acoustics to study the ocean (e.g., Howe et al. 2019). And now their arsenal includes acoustic tools attached to the newest recruits in oceanographic research—marine animals. These approaches belong to the field of **acoustical oceanography**, the application of sound and its properties to study the ocean.

Acoustical oceanography applies to a broad range of oceanographic disciplines. It's no surprise. In a medium largely opaque to light, sound offers the ability to "see" what otherwise might remain hidden. Listen closely to the sounds around you. Experience the world as whales and dolphins do. Enter the world of acoustical oceanography.

FIGURE 5.3 Scientists believe the songs of humpback whales—sung by the males—are meant to attract a future romantic partner, but it's okay if we enjoy them, too.

Credit: NOAA/Anne Smrcina

5.1 Anatomy of a Sound Wave

Before we dive into how oceanographers use sound to study the ocean, we need to know something about the nature of sound itself. Dictionaries define **sound** as vibrations that you can hear. But we must take a broader view because many underwater sounds fall outside the range of human hearing. Sounds originate through vibrations of an object that travel as pressure waves through air, water, and even solids. Thus, we may define sound as "a mechanical disturbance that propagates as a pressure wave through an elastic medium" (modified from britannica.com).

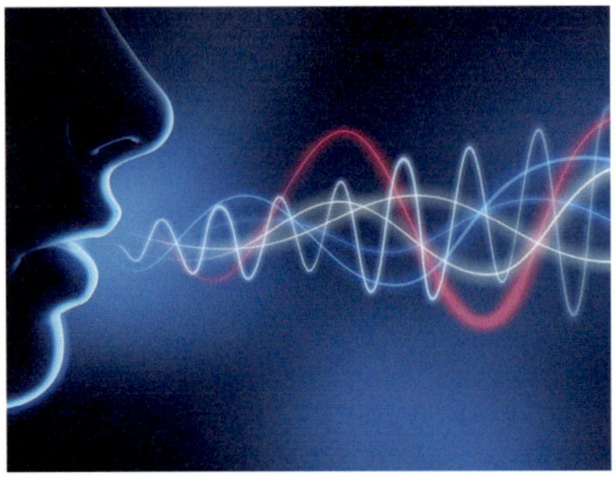

FIGURE 5.4 Vibrations in your voicebox transfer kinetic energy to air molecules, which then move through the air as sound waves.

Credit: istock/peterschreiber.media

Sound waves move through the air by compression and expansion—more commonly referred to as **rarefaction**—of the air molecules. The completion of one compression-expansion motion of air molecules is termed a **cycle**, or oscillation. We can represent this cycle as a **sine wave**, a graphical representation of the sine function, a concept from trigonometry. A sine wave has a **sinusoidal** appearance—that is, it's a line with regularly spaced arches. The top of the arch represents the **crest** of the wave and the bottom of the arch the **trough**. The horizontal distance between successive

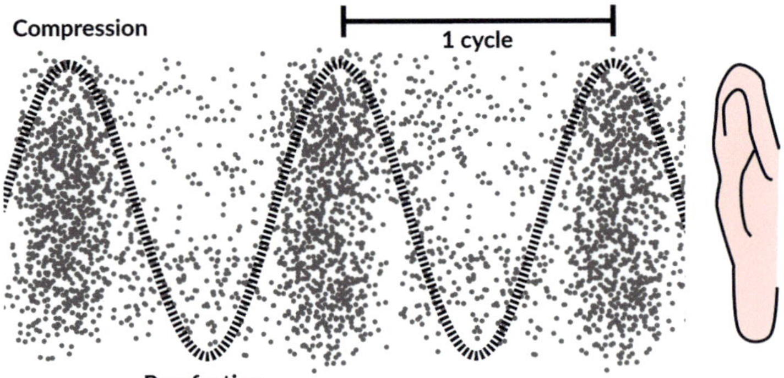

FIGURE 5.5 Sound energy alternately compresses and expands air molecules as it travels outward from its source.

Credit: Wikipedia/Pluke; modified by Chamberlin

crests (or troughs) is called the **wavelength**. The height of the crest above the undisturbed baseline (the zero line) is called the **amplitude**. With sound waves, amplitude determines the "loudness," or the intensity of the sound we perceive. Amplitude is a measure of the amount of energy a wave carries. Higher-amplitude waves sound louder. Lower-amplitude waves sound softer. Slam a key on a piano, and you get a high-amplitude sound (i.e., loud). Touch the same key gently, and you get a low-amplitude sound (i.e., soft).

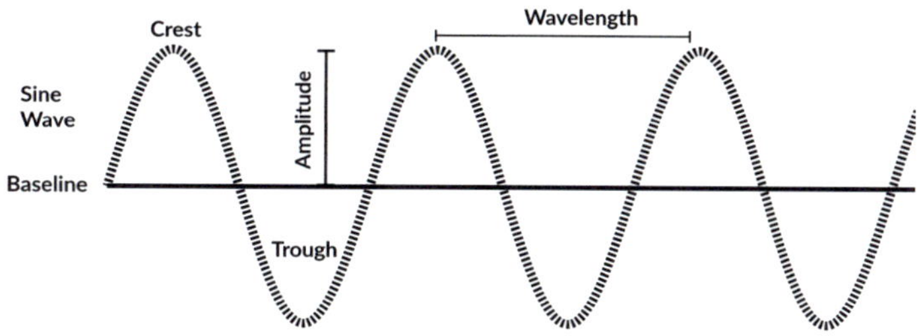

FIGURE 5.6 Parts of a sound wave.

Credit: Wikipedia/Pluke; modified by Chamberlin

Sound waves travel in a forward direction. Thus, they are called **progressive waves**, waves that move outward from their source. Progressive waves represent the transfer of energy between molecules in the medium through which they travel. The ocean swell you observe from a pier at the ocean is another kind of progressive wave.

The forward movement of progressive waves adds a time element to their description. We define **wave speed** as the distance a wave travels over a given period of time. Wave speed is related to two other

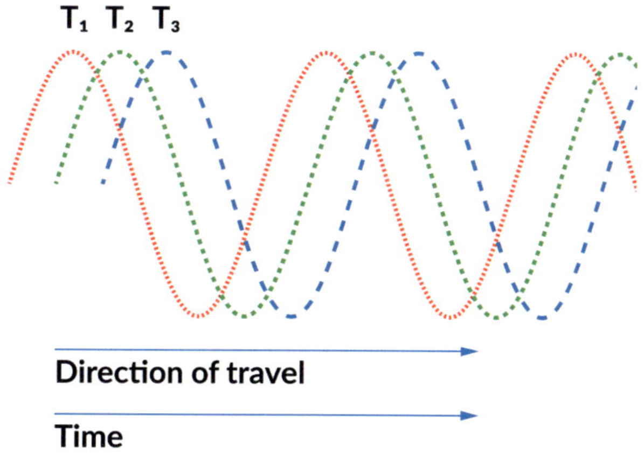

FIGURE 5.7 Progressive waves travel outward from their source.

Credit: Chamberlin

FIGURE 5.8

DJs spin records which cause vibrations that are transformed into radio waves and transmitted at different frequencies. WFC-AM and WKYS-FM radio in Washington DC in December 1977.

Credit: Library of Congress

wave variables. **Wave period** represents the interval of time between successive crests. **Wave frequency** indicates the number of cycles in a fixed interval of time (usually a second). You may be familiar with different frequencies of radio waves used to transmit sounds via radio, television, smartphones, and other devices.

Frequency is generally reported in units of **hertz** (Hz; 1 cycle per second) or **kilohertz** (kHz; 1,000 cycles per second). Human hearing typically ranges from 20 hertz to 20 kilohertz (but this varies by individual); this range is called **audible** (or sonic) sound. At the low end of the range (low-frequency sounds), you'll find a bass drum or the rumble of a truck. At the high end of the range (high-frequency sounds), you'll find crickets and light rain. Scientists define sounds outside the range of human hearing, such as those produced by humpback whales and other animals, as **infrasonic** or subsonic (frequencies less than 20 Hz) or **ultrasonic** (frequencies greater than 20 kHz).

FIGURE 5.9

AM radio modifies the amplitude of the signal while FM radio alters the frequency to transmit sound information.

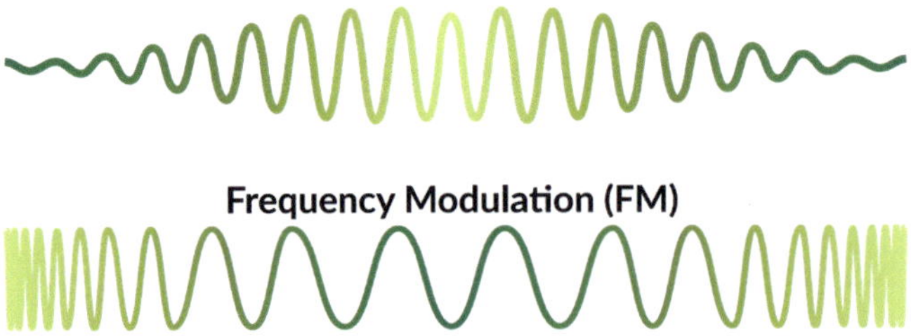

Credit: istock/petrroudny; modified by Chamberlin

5.2 Factors Affecting the Speed of Sound

The speed of sound in a fluid such as water or air depends on the fluid's density; the denser the fluid, the faster sound travels. Because seawater is 800 times more dense than air, sound travels about 4.4 times faster in the ocean versus in the air, about 4,921 feet per second (1,500 meters per second [m/s]) versus 1,115 feet per second (340 m/s), respectively. But these speeds are not constants. Factors that affect the density of seawater (or air) can also cause variations in the speed of sound.

When we increase the temperature, salt concentration (salinity), or pressure of seawater, we increase the speed of sound. Now, in a typical ocean, temperature is highest at the surface, then rapidly decreases and stays relatively constant with increasing depth. So, based on temperature alone, the speed of sound would be highest at the surface, then slow down and remain constant with greater depth. However, salinity (usually) and pressure (always) increase with depth. These two factors alone would cause sound to speed up. The opposing effects of the three combined—

FIGURE 5.10 The speed of sound in seawater is more than four times the speed of sound in air.

Credit: Shaw and Chamberlin

FIGURE 5.11

Generalized profile of the speed of sound in seawater. Note the sound minimum layer at approximately 1,000 meters.

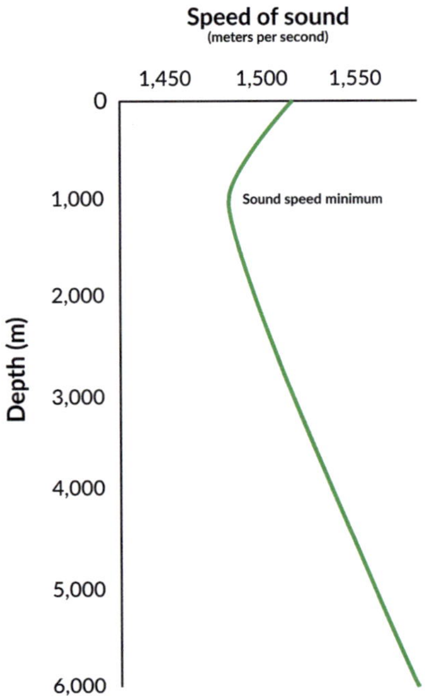

Credit: Chamberlin

temperature slowing sound and salinity and pressure speeding it up—create a curved sound speed profile in a typical ocean. Near the surface the speed of sound is fast (because temperature is high), but as temperature decreases, the speed of sound slows down. At around 3,281 feet (1,000 m), temperature remains constant, and the effects of increasing salinity and pressure cause sound to speed up. The result is that the sound speed reaches a minimum at around 3,281 feet (1,000 m) in the ocean. Above and below 3,281 feet (1,000 m), the speed of sound increases. Oceanographers refer to this depth as the **sound minimum layer**.

FIGURE 5.12

Sound velocity profiles at six locations in the world ocean (blue and yellow solid lines) compared to model estimates based on XBTs (black dashed) and the World Ocean Atlas 2009 (red dashed). Measured and modeled estimates mostly agree, except in the upper 600 m of the Mariana Trench and Molloy Hole. See Bongiovanni et al. 2021.

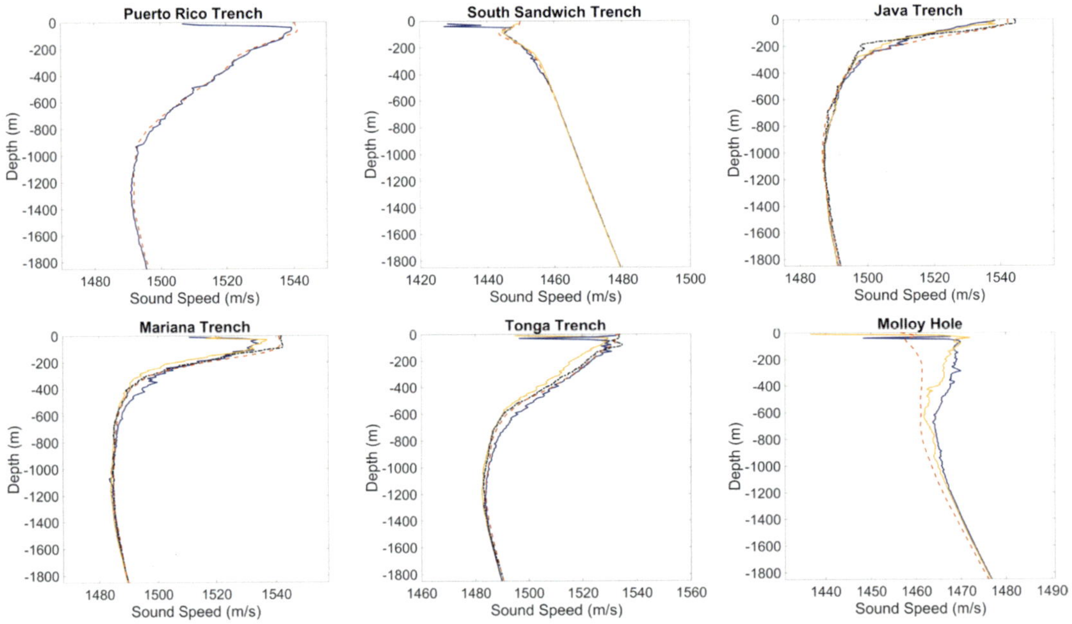

Credit: Bongiovanni et al. 2021 CC BY 4.0

5.3 The SOFAR Channel and SOSUS

The subsurface minimum in the speed of sound has enormous implications for animals (and humans) who use sound for communication in the ocean. The sound minimum layer has an unusual effect on sound traveling at these depths. It causes sound waves that move shallower or deeper than the sound minimum layer to bend back into the layer. This bending of the waves—called **wave refraction**—prevents the sound waves from leaving the sound minimum layer. This "channeling" of sound waves allows them to travel hundreds to thousands of miles through the deep ocean. Discovery of this property led to recognition that it could be helpful in underwater communications. The US Navy equipped life rafts with small underwater explosives that could be detonated in the channel to assist ships in finding sailors or pilots lost at sea during World War II. Henceforth, the sound minimum layer became the **Sound Fixing and Ranging, or SOFAR, channel**. Because the SOFAR channel is so efficient at propagating sound over long distances, scientists speculate that marine mammals, such as blue whales, use it to communicate (Tsuchiya et al. 2004).

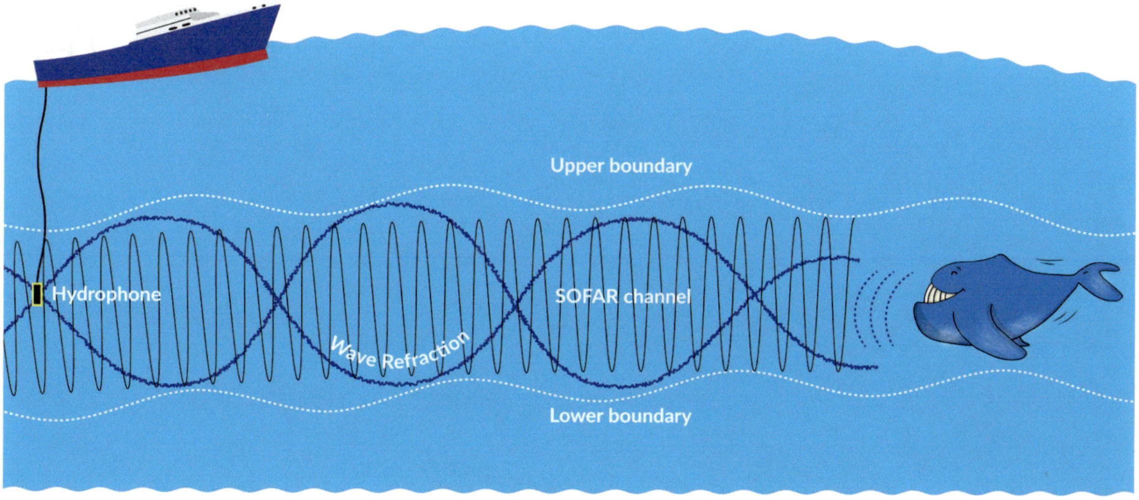

FIGURE 5.13 Refraction of sound waves in the sound minimum layer prevents them from traveling above or below the SOFAR channel.

Credit: istock.com/iCexpert; Shaw and Chamberlin

FIGURE 5.14
Blue whales and other sound-producing species may use the SOFAR channel for long-distance communication.

Credit: Shaw and Chamberlin

In the 1950s the US Navy installed arrays of hydrophones to detect noisy submarines and surface vessels in the SOFAR channel. The project, known as the **Sound Surveillance System, or SOSUS**, successfully detected Soviet submarines until the Soviets learned about SOSUS and started making their submarines quieter. Declassification of the SOSUS recordings in the 1960s gave researchers a golden opportunity to study underwater sounds, including the mysterious ones that puzzled hydrophone specialists like Frank Watlington.

FIGURE 5.15
The US Navy continues to deploy and maintain hydrophone arrays on the seafloor. This one—near Barking Sands, Kauai, Hawaii—is used for surveillance, training, and marine mammal monitoring.

Credit: US Navy/Ben Reed

5.4 Echolocation and Sonar

As scientists developed tools for listening to underwater sounds, they also worked to perfect tools using "echoes" for finding underwater objects such as icebergs and submarines, a process known as **echolocation**. The principle is familiar to anyone who has created their own echo against a wall. You make a loud noise. A split second later, an echo comes back to you.

Credit: Shaw and Chamberlin

FIGURE 5.16 Dolphins use echolocation to find and identify objects, food, and other animals in their environment.

Echolocation—finding an object using sound—originated in animals, such as bats, birds, and marine mammals, who use echolocation to find food and sonically visualize their environment (Brinklov et al. 2013; Surlykke and Nachtigall 2014). Humans can use echolocation to determine water depth. By accurately measuring the time between the start of the sound and detection of the reflected echo, the distance to the seafloor can be estimated. Of course, the distance traveled by the reflected sound

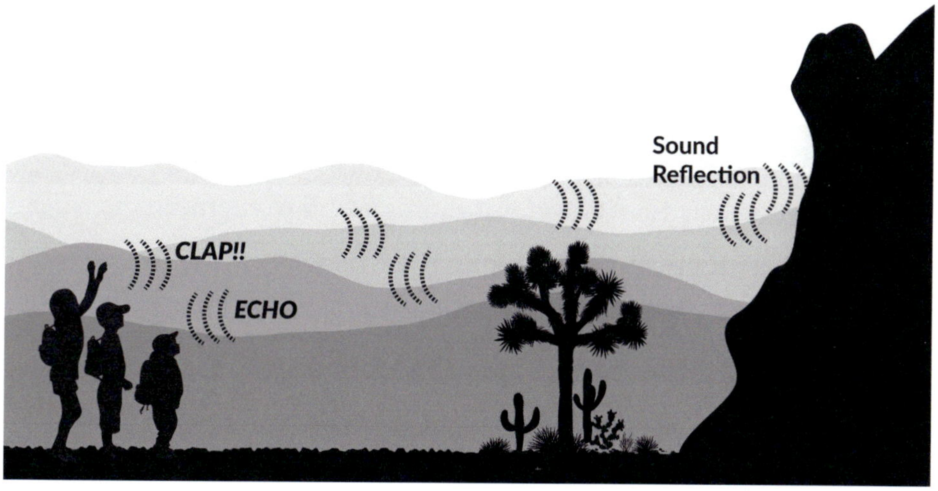

Credit: istock/Toltemara; istock/syntika; and Chamberlin

FIGURE 5.17 Making your own echoes against a wall or rock face simulates the principles of echolocation.

FIGURE 5.18
A Navy sonar technician adjusts the sonar signal aboard the guided-missile destroyer USS Lake Erie.

Credit: US Navy/Lucas T. Hans

waves is twice the distance between the source and the seafloor. Knowing the time gives you the total distance traveled by the sound; dividing by two gives you the one-way distance to the seafloor, the water depth.

Today we know echolocation devices as **sonar**, an acronym for sound navigation and ranging. If you watch submarine movies, you're probably familiar with the ping of **active sonar**. This range-finding tool produces a pulse of sound to locate an object through its echo. Modern sonar can employ multiple signals of various frequencies such that, at close range, the sonic image, when displayed in three dimensions, looks almost as detailed as a photograph.

5.5 Types of Sonar and Their Applications

Active sonar tools support a field of science known as **geophysics**, the study of solid Earth through observations of its physical properties. Marine geophysics, exploration of solid Earth beneath the ocean, has contributed significantly to our understanding of plate tectonics, formation of petroleum reserves, and much more. Of course, sonar also ensures safe navigation for anyone who ventures forth in a vessel in shallow waters and provides fishers an indispensable tool for finding fish. In that regard, sonar has evolved into a tool with a surprising variety of applications.

5.5.1 Fathometers

One of the earliest applications of sonar was determining the depth of the seafloor. Before sonar—since at least the sixth century BCE—mariners relied on a device called a **sounding line** (or lead line), which consisted of rope (or wire) and a lead weight on which a piece of candle wax (made from animal fat) was placed (e.g., Hawley 1931; Oleson 2000). The Challenger expedition carried miles of rope to sound for depth. The line and weight were lowered over the side of a ship while keeping track of the arm-lengths of rope deployed, a unit of measurement that came to be known as a **fathom**, equal to 6 feet (1.8 m). Of course, if you lowered hundreds (or even thousands) of feet of rope, you had to raise it to the surface to retrieve it. If you've ever worked with ropes in the gym, you can imagine the workout from raising a few miles of rope from the seafloor! Once at the surface, the weight was inspected for the presence of sediments in the wax, a sign that it had reached the bottom.

FIGURE 5.19 Mapping the depths using a sounding line aboard the US Coast and Geodetic survey ship *Explorer* in the 1930–40s (Adams 1942, 387).

Credit: NOAA

The term "fathom" originated in the 800s CE as an expression for "outstretched arms." A fathom of depth approximates the distance of a person's outstretched arms, about six feet. It also describes the process of understanding something complex or mysterious.

FIGURE 5.20
A fathometer sends a pulse of sound to the seafloor and listens for its reflection from the seafloor. The time interval from ship to bottom and back provides an estimate of depth, i.e., 0.5 × time × speed of sound.

Credit: Shaw and Chamberlin

Because **soundings** refer to depth measurements, oceanographers adopted echosounding to refer to depth measurements using sound. However, the first device, invented by German physicist Alexander Behm (1880–1952), set out to detect icebergs following the sinking of the *Titanic* in 1912. Eventually, these instruments were turned downward to measure seafloor depth. They came to be known as **fathometers**, a word invented to describe the device patented by Canadian physicist Reginald Fessenden (1866–1932). Swarthmore College physicist Harvey Hayes (1878–1968) developed the first working fathometer in 1922. This instrument was deployed that same year off the coast of Southern California. This device helped to produce the first map of seafloor depths—what is known as a **bathymetric chart**. Comparisons of

FIGURE 5.21
Closeup of bathymetric chart 5101A produced in 1939 by the Coast and Geodetic Survey from measurements made by sounding lines and fathometers.

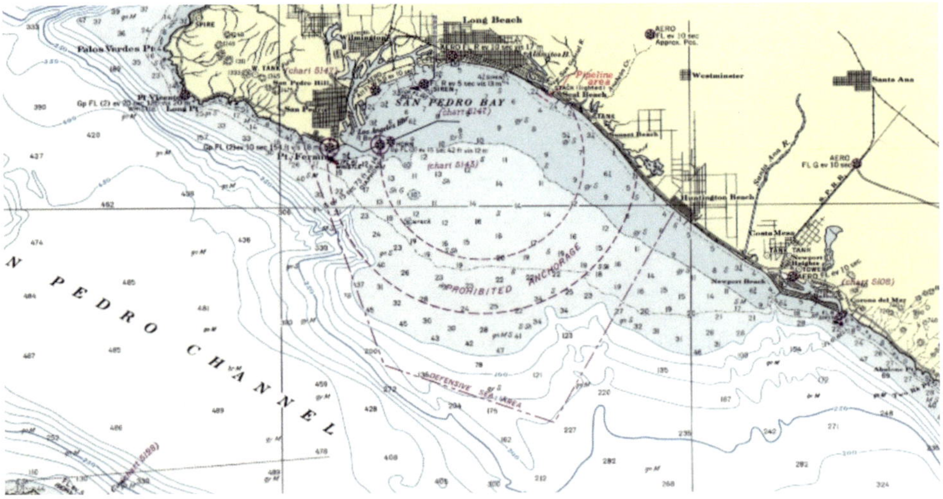

Credit: NOAA

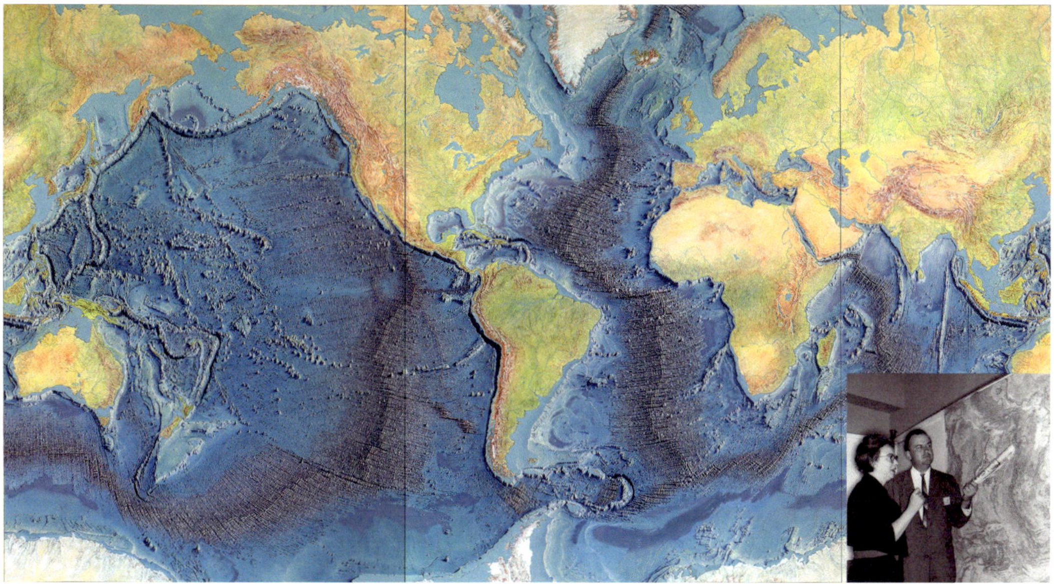

FIGURE 5.22 The World Ocean Floor map painted in 1977 by Heinrich Berann under the direction of by Bruce Heezen and Marie Tharp (inset) represents the pinnacle of fathometer-based ocean bathymetry.

Credit: Library of Congress (map); Fiona (inset) CC BY 2.0

fathometer readings and sounding lines made by the US Coast and Geodetic Survey proved so favorable that the navy placed the instruments on all its ships soon afterward (e.g., Adams 1942; Hersey 1977; D'Amico and Pittenger 2009).

In 1957 echosounder data compiled by American oceanographers Bruce Heezen (1924–1977) and Marie Tharp (1920–2006) revealed the presence of a continuous range of undersea mountains running down the middle of the North Atlantic (e.g., Heezen et al. 1959). The Heezen-Tharp bathymetric map of the North Atlantic uncovered the Mid-Atlantic Ridge, an early piece of evidence supporting plate tectonics theory (see Chapter 8). Nowadays, powerful and much-improved fathometers can be found on nearly every vessel afloat.

5.5.2 The Deep Scattering Layer

Almost immediately after fathometers became available to the general public, vessel operators and fishers put them to use. But the instruments produced strange signals, ghost reflections, and other aberrations in the soundings. Military sonar operators noted the appearance of "false bottoms" in their fathometer displays. The phenomenon became known

FIGURE 5.23 The mysterious "false bottoms" in fathometer readings turned out to be small organisms and fishes, which migrate up and down in the water column.

Credit: Shaw and Chamberlin

as the **deep scattering layer** (or sound scattering layer), a depth layer at which sound waves were reflected in different directions. More curious, the layer moved up and down at different times of day. Scripps biological oceanographer Martin Johnson (1893–1984) was among the first to propose that the layer was composed of organisms. The curious motions of the layer resulted from **diel vertical migration**, the daily movement of organisms up and down in the water column. An experiment on a research cruise in 1945 confirmed his hypothesis (Johnson 1948).

Modern instruments and automated processes have improved tools for observing organisms acoustically. Acoustic oceanographer D. Van Holliday (1940–2010) and biological oceanographer Rick Pieper (*b.* no data) spent decades studying zooplankton using acoustical methods (e.g.,

FIGURE 5.24 Plankton, krill, and fish make up the deep scattering layer, which rises to the surface at night.

Credit: istock/izanbar

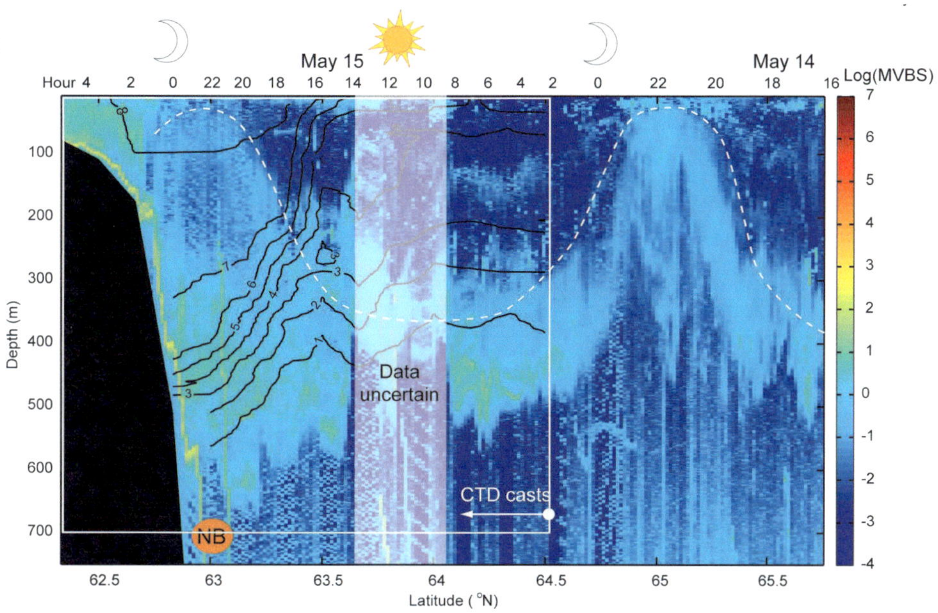

FIGURE 5.25 The thick light blue layer—the deep scattering layer as revealed by acoustic doppler instruments—rises at night and descends during the daytime in the Norwegian Sea. It's the distinctive signature of vertically migrating animals.

Credit: Cisewski et al. 2021 CC BY 4.0

Holliday and Pieper 1995). The latest generation of acoustic instruments can now be deployed on moorings (De Robertis et al. 2017), towfish (Wiebe et al. 2002), remotely operated vehicles (ROVs; Dunlop et al. 2020), and autonomous underwater vehicles ((AVUs; Dunlop et al. 2018). These instruments have yielded great insights into the deep scattering layer (e.g., Fornshell and Tesei 2013; Receveur et al. 2020; Cisewski et al. 2021).

5.5.3 Multibeam Echosounders

Multibeam echosounders embody the idea that multiple echoes are better than one. Developed in the 1960s by the US Navy and put into commercial use in the 1970s, multibeam echosounders use multiple sound sources and receivers to obtain seafloor depths at many points simultaneously. Think of a spray nozzle whose liquid emanates from a point and spreads like a fan. That's the shape of the

FIGURE 5.26 Multibeam echosounders emit a fan-shaped "spray" of sound pulses to image the seafloor.

Credit: istock/VladmirFLoyd

FIGURE 5.27 Like mowing a lawn, ships conduct multibeam bathymetry moving back and forth across the seafloor.

Credit*: *istock/yunava1

beam transmitted by multibeam echosounders. This fan-shaped sound bounces echoes off a strip of the seafloor, and the return echoes are captured by receivers on the hull of the ship. By capturing multiple strips per second and digitally stitching them together, a three-dimensional picture of the seafloor can be obtained. To map large areas, oceanographers use the **mowing the lawn technique**—a method where the instrument travels back and forth across the seafloor similar to mowing a lawn. A ship traveling at a moderate speed can map hundreds of square miles a day in this way.

Multibeam bathymetry—mapping seafloor depths using multibeam echosounders—builds on work that began as far back as 1870. Albert I, Prince of Monaco (1848–1922), a self-taught oceanographer, sought to produce the most comprehensive maps of the ocean the world had ever

FIGURE 5.28 Multibeam bathymetry produces 3D images of the seafloor. In this example from the Great Barrier Reef in Australia, an ancient river bed was revealed.

***Credit*: AUS SEABED/*Rob Beaman* CC BY 4.0**

seen. Prince Albert organized a committee to compile all available data sources on the ocean's depths. In May 1905 they published the first edition of the General Bathymetric Chart of the Oceans, or GEBCO. Still available in modern times—albeit in digital form—GEBCO remains the primary source of data on global seafloor bathymetry. (See also Weatherall et al. 2015).

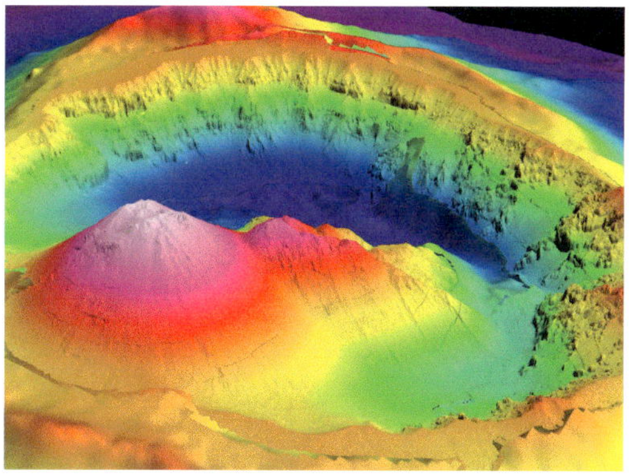

FIGURE 5.29
A 3D image of the Brothers Volcano near New Zealand imaged by an AUV equipped with multibeam sonar.

***Credit*: NOAA CC BY 2.0**

Unfortunately, modern bathymetric maps of the deep ocean lack the spatial resolution to resolve features smaller than three miles (5 km) across. Imagine a stadium (such as Pasadena's Rose Bowl) filled with people standing shoulder to shoulder. Drop an enormous sheet on top of the people, and you would see a surface with bumps and depressions according to the heights of the people. To an echosounder operating in the ocean far above, all those bumps and depressions would condense to one depth. None of the details would be visible. That's the challenge of mapping the ocean.

To address this shortcoming, GEBCO recently launched the Seabed 2030 Project to produce the highest-resolution map of the seafloor possible by 2030. To obtain a more detailed picture, scientists will require hundreds of multibeam-equipped AUVs and ROVs that can get closer to the bottom but that travel at much slower speeds. As a compromise, different depths will be mapped at different resolutions. They admit "that the goal of seeing the world ocean completely mapped by 2030 is a difficult one," but they hope their efforts will inspire others to join them (Mayer et al. 2018).

5.5.4 Side-Scan Sonar

When oceanographers require detailed images of the seafloor, they reach for their **side-scan sonar**. Transmitting beams of sound from both sides of the instrument, which is most often towed near the seafloor, side-scan sonar provides images of "near photographic quality." The high resolution and contrast of side-scan sonar images, called **sonographs**, make this technology ideal for identifying human-made objects on the seafloor, such as ships, planes, mines, pipelines, and similar structures (e.g., Kenny et al. 2003; Blondell 2009; Savini 2011; Luchetti et al. 2018; Zhang et al. 2021). Side-scan sonar works like the medical ultrasound instruments used to generate sonographs of internal organs. Its success in finding shipwrecks spurred its application in oceanographic research, especially in marine geology and geophysics. Side-scan sonar has even proven useful for tracking marine organisms (e.g., Grothues et al. 2017; Guzman and Condit 2017).

FIGURE 5.30 Side-scan sonar works much like ultrasound technology used to image tissues and organs in people and pets.

Credit: istock/RossHelen

FIGURE 5.31 Oceanographers prepare the KATFISH™, a towfish with synthetic aperture sonar, the latest generation of side-scan sonar instruments.

Credit: NOAA

While side-scan sonars produce highly detailed images, they're two-dimensional. They lack information on depth or the vertical relief of an object. To overcome this limitation, oceanographers use multibeam echosounders alongside side-scan sonar or platforms that integrate both technologies (e.g., Shang et al. 2019). With computer software, they can overlay the multibeam bathymetry on top of the side-scan image and obtain a sonograph with bathymetry.

FIGURE 5.32 KATFISH™ SAS image of the USS Bass, a "retired" submarine on the seafloor off Rhode Island.

Credit: NOAA, ThayerMahan, Inc. and Kraken Robotics

Synthetic aperture sonar (SAS)–such as the KATFISH™ SAS–represents a new-and-improved version of side-scan sonar. Synthetic aperture sonar systems produce a much higher resolution image by averaging multiple samples (i.e., multiple return echoes) and combining them (e.g., Hansen 2019). The result is a kind of ultra-high-definition sonograph. Smile, fishes!

The search-and-recovery group Adventures with Purpose uses side-scan sonar to find vehicles and missing persons in lakes and other bodies of water. Learn more on their website.

5.6 Seismic Methods

We now turn our attention to sound used for observing features beneath the seafloor. Though not generally classified as echosounding, these techniques use sound waves to infer characteristics of the **subseafloor**, the region greater than 4.9 feet (1.5 m) beneath the seafloor surface (e.g., Walsh et al. 2016). These tools revolutionized exploration for petroleum. In modern times, these tools have helped us to understand plate tectonics and other geological and oceanographic processes. You'll enjoy this section if you like explosives (or explosive sounds, at least).

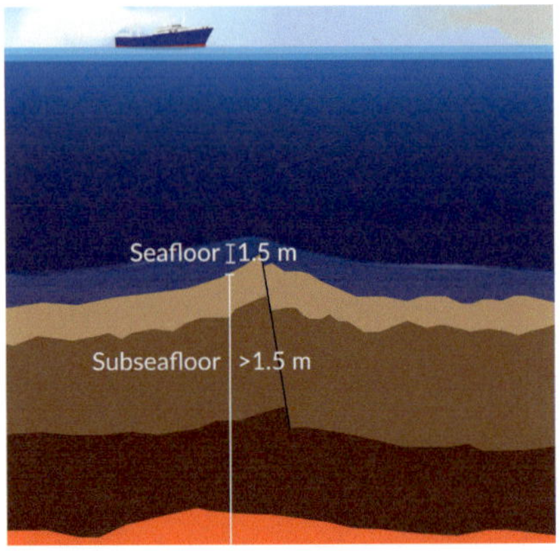

FIGURE 5.33 Seismic methods reveal the features of the subseafloor, the region below the seafloor.

Credit: Chamberlin

5.6.1 Seismic Profiling Overview

Seismic profiling uses artificially produced seismic waves to map the three-dimensional structure of Earth's crust. The technique originates in studies of earthquakes, a field of science known as **seismology**. When a segment of Earth's crust under high stress ruptures, the crust moves, and an earthquake occurs. We feel an earthquake as a series of elastic vibrations called **seismic waves**, motions of the ground caused by the movement of energy through Earth's crust. These ground motions can be detected using an instrument called a **seismometer**, a device for detecting and recording the three-dimensional movements of Earth's crust during the passage of seismic waves.

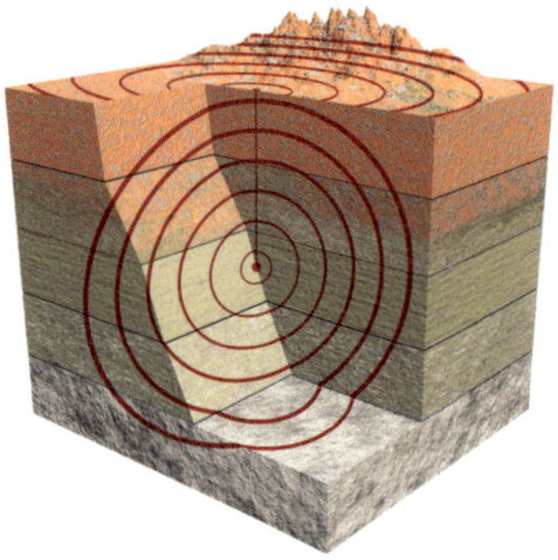

FIGURE 5.34 Earthquakes generate seismic waves that travel through Earth's interior and on its surface.

Credit: istock/Naeblys

In 1849 Irish engineer Robert Mallet (1810–1881), the "founder of seismology" (Dean 1991), struck upon the idea of generating artificial earthquakes to study how seismic waves travel through Earth's crust. In a series of experiments in 1849 and 1850, he generated underground explosions using gunpowder. He then measured the speed of the resultant seismic waves. He discovered seismic waves travel faster in granite than sand (Mallet 1851). Mallet's work established the field of **artificial seismology**, the use of human-generated seismic waves for studying the Earth's interior. Subsequent work built on Mallet's work using different explosives and improved methods for detecting seismic waves (Weatherby 1940). By the turn of twentieth century, seismologists had worked out the principles of **seismic reflection profiling**, the detection of the echoes of seismic waves that bounce off subsurface rock layers. It's like echosounding, except the echoes come from beneath Earth's surface (Hedström and Kollert 1949). By the turn of the century, with the Industrial Revolution in full swing, scientists could use

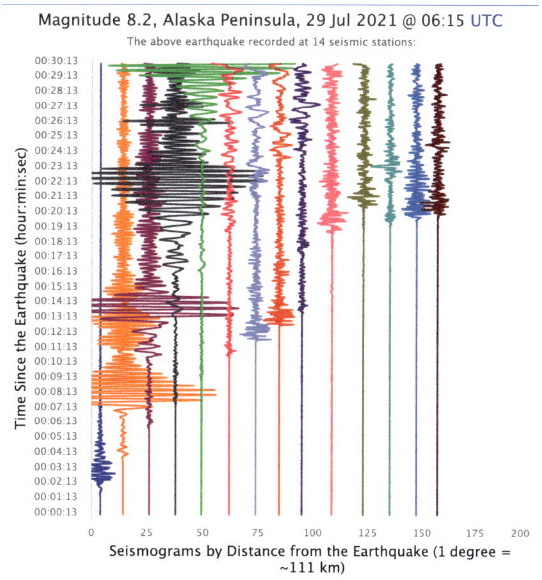

Credit: Incorporated Research Institutions for Seismology

FIGURE 5.35 Digital seismograms at different seismograph stations following the magnitude 8.2 earthquake off southern Alaska in 2021.

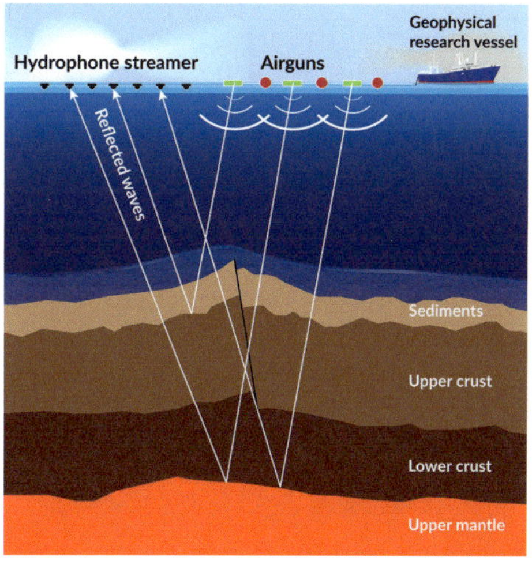

Credit: Chamberlin

FIGURE 5.36 Schematic of seismic reflection profiling of the subseafloor.

FIGURE 5.37 Seismic prospecting still involves the use of explosives, but much smaller (and safer) blasts are used than in the past.

Credit: istock/westphalia

seismic methods for petroleum and mineral exploration, a field of applied geophysics known as **seismic prospecting**.

Seismic profiling requires two key ingredients: a source to generate seismic waves and a means to detect the returning waves. Dynamite was once used to generate seismic waves on land and at sea. And while explosives remain in use on land, seismologists more commonly drop heavy weights or employ **thumper trucks**, vehicles with devices that pound the ground to create vibrations. Return echoes may be detected by **geophones**—microphones that listen for the sound vibrations caused by seismic waves on land. At sea, loud booms are produced through the sudden release of compressed air using a device called a **seismic airgun**. To detect the sounds, seismologists use hydrophones or **ocean bottom seismometers**, devices anchored on the seafloor that measure seismic

FIGURE 5.38 Thumper truck assisting a seismic survey in Saudi Arabia.

Credit: istock/Elena Abrosimova

waves. The hydrophones or ocean bottom seismometers detect the acoustic reflections—which arrive at different times depending on the depth and density of the layers—to obtain a visual image of the rock layers beneath the seafloor.

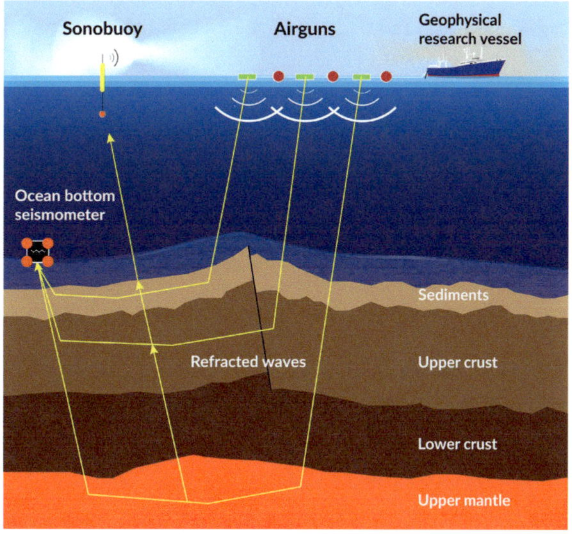

FIGURE 5.39 Schematic of seismic refraction profiling of the subseafloor.

Credit: Chamberlin

Seismic waves also undergo **refraction**—the bending of a wave front. Refraction in seismic waves occurs in response to the nature of the rock through which they travel; some rock layers permit waves to move faster, while others slow the wave down. Because seismic waves from deeper layers in Earth's interior return to the surface at a greater distance from their source than refracted waves from shallower layers, the speed of return of seismic waves reveals something of the nature of the rocks in the interior. Based on these principles, **seismic refraction profiling**—the use of refracted seismic waves for determining shallow surface geologic structures—was developed.

FIGURE 5.40 Airguns (attached to orange floats) towed behind a ship as part of a seismic study of the Lau Basin near Fiji.

Credit: GEOMAR/Philipp Brandl CC BY 4.0

5.6.2 Seismic Profiling of the Subseafloor

Both seismic reflection and seismic refraction profiling rely on the subsurface penetration and reflection of sound waves produced by a loud acoustic source, e.g., airguns. Where seismic reflection and seismic refraction differ is in the placement of detectors. Seismic reflection uses a string of airguns and hydrophones, called a **streamer**, towed behind a ship. Streamers may be several kilometers long and hold thousands of hydrophones. Seismic refraction uses seismometers and hydrophones at the ocean bottom to detect the refracted wave from surface-towed airguns. However, seismic refraction requires much greater distances to characterize the subsurface layers. So detectors are placed further from the source, perhaps as much as 19 to 25 miles away (30–40 km).

To obtain a profile, airguns are triggered at regular intervals (every 10–12 seconds) while the vessel is underway (at about 5 mph). The pulse of sound reflects off the seafloor (to 6 miles deep) and echoes back to the streamer. Sound waves are reflected and refracted by layers of sediments and rock beneath the seafloor. The wave arrival times and other characteristics can be combined using computer visualization tools to provide a **seismic reflection (or refraction) profile**, a two- or three-dimensional image of the subseafloor. These graphs reveal the thickness of different layers, buried subsurface features such as salt domes or seamounts, discontinuities and faults, and other features within the layers that make them valuable for oil and mineral

FIGURE 5.41
The GeoEel hydrophone streamer looks like a green garden hose, but it's packed with multichannel digital hydrophones that record sounds from a seismic source.

Credit: US Geological Survey

FIGURE 5.42 Oceanographers reel out the GeoEel hydrophone streamer as part of a seismic reflection study off the coast of Puerto Rico.

Credit: USGS

exploration and scientific research. (See Crutchley and Kopp 2018 for a recent review of these methods.) As seismic methods have revealed, a whole other world exists beneath our feet.

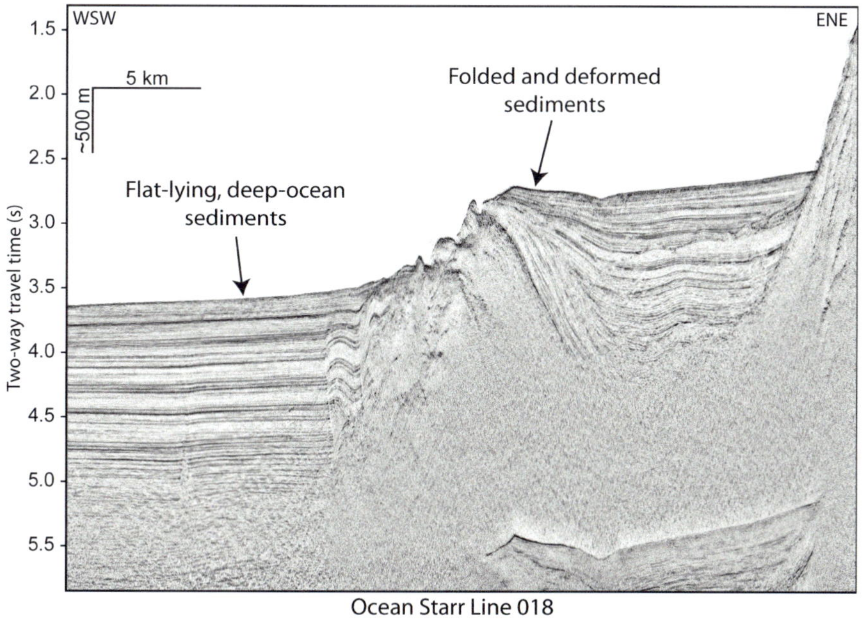

FIGURE 5.43 Seismic reflection profile of the subseafloor off Glacier Bay, Alaska.

Credit: US Geological Survey

FIGURE 5.44 Two vessels carry out seismic reflection surveys at dusk in a search for subseafloor gas and oil in the Gulf of Mexico. Magnificent sunsets are one perks of working on ships at sea.

Credit: istock/landbysea

5.6.3 Seismic Oceanography

When seismic reflection profiling became a common tool for mapping the geology of the subseafloor, geophysicists noticed reflections in the water column between the ship and the seafloor. They paid them no attention. After all, they were interested in what was beneath the seafloor. Why care about "nuisance" false echoes (Buffett and Carbonell 2011)? As sometimes happens in science, those "echoes" are important.

FIGURE 5.45 Like an inflated inner tube, less dense layers of water float on top of more dense ones.

Credit: istock/Rawpixel

If you swim in a lake during summer, you know that the surface can be quite warm. But deeper—where you stick your feet—it can be downright freezing (at least it feels that way). Lakes and the ocean form layers. Because warm water is less dense than cold water, the warm layer floats on top of the cold layer. The interface between these layers is called a **thermocline**. It represents the boundary between different density layers in a lake or the ocean. As it turns out, those density layers reflect low-frequency sound. And those density layers are the echoes oceanographers saw in the water column during seismic reflection profiles.

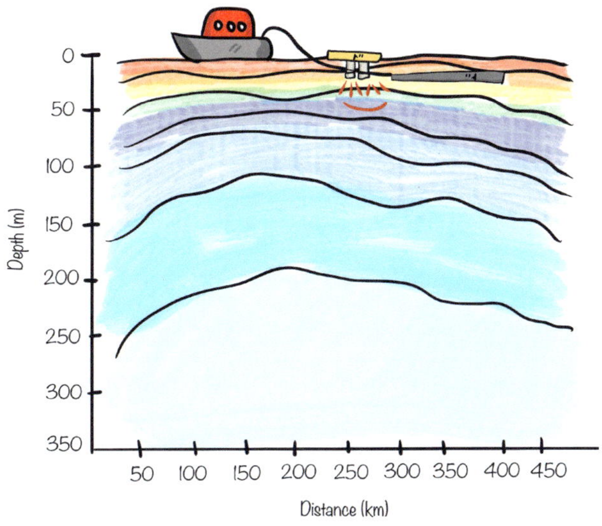

FIGURE 5.46 Seismic reflection profiling reveals thermoclines, boundaries between waters of different temperature.

Credit: Shaw and Chamberlin

In 1988, French oceanographers Joseph Gonella and Dominique Michon took an interest in those reflections. They were soon followed by American oceanographers Joseph Phillips and Donald Dean. While

FIGURE 5.47 Crewmembers aboard the US Naval Ship *Pathfinder* prepare to launch an XBT in the Gulf of Mexico.

Credit: US Navy/Bobby Dixon

they took different approaches, both presented compelling evidence thermoclines reflected sound. Their work deserved attention. As Phillips and Dean (1991) put it, "The past reluctance to use the multichannel reflection method for investigating water mass structure appears to have been unfortunate."

Nevertheless, in 2003, publishing in the prestigious journal *Science*, Holbrook et al. presented "acoustic images" of thermoclines, fronts, and water masses in the North Atlantic. Anywhere density differences occurred, their images showed a reflection. To verify the seismic observations, the scientists deployed **expendable bathythermographs** (also known as XBTs). These disposable temperature recorders transmit data through a wire to the ship. Onboard chart recorders or computers record the temperature of the water column as the XBT descends. The data from the two approaches matched, demonstrating the utility of seismic reflection profiling for imaging the water column.

FIGURE 5.48

Though perhaps not obvious to the untrained eye, the lines and wiggles on the top chart—obtained from seismic reflection profiling of the water column—corresponds to the temperature data in the bottom chart, obtained using data from XBT and CTD profiles.

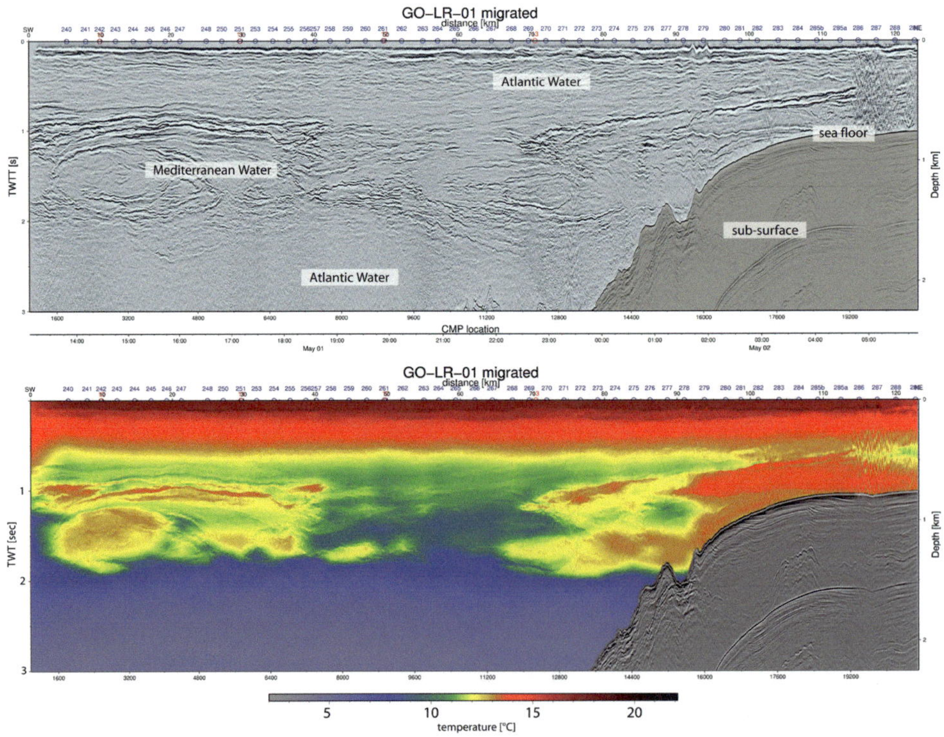

Credit: Papenberg et al. (2010)/GEOMAR CC BY 4.0

Persistence pays off in science, especially when you know your data and their interpretation are correct. Seismic oceanography has emerged as an important tool for understanding the ocean.

Upon publication of the Holbrook paper, the oceanographic community moved quickly to "see" what was previously "unseen." Several papers by other researchers followed (e.g., Nandi et al. 2004; Nakamura et al. 2006; Biescas et al. 2008; Ruddick et al. 2009; Papenberg et al. 2010; Holbrook et al. 2013; Gunn et al. 2020). Thus began the field of **seismic oceanography**, the application of seismic reflection profiling to studies of water column structure and dynamics.

Seismic methods promise to make large-scale structures such as fronts and eddies visible on an unprecedented scale. It's like that heist film where the thieves blow smoke into the room to light up the laser motion sensors. As Gunn et al. (2020) put it, "Seismic reflection surveying has a hitherto unsurpassed ability to resolve thermohaline structures on spatial scales of tens of meters to hundreds of kilometers and on temporal scales of minutes to days." That means oceanographers can now see the physical structure of the ocean as never before.

5.7 Marine Biotelemetry and Biologging

Oceanographers have also developed acoustic tools for observing individual organisms. These tools have brought insights into marine animal behavior not possible just a few decades ago (e.g., Block et al. 2016). Two types of animal tracking technologies have emerged in recent decades: (1) **biotelemetry**, the tracking of organisms using devices that continuously transmit signals to a receiver; and (2) **biologging**, the recording of data by devices attached to or embedded within an organism (e.g., Bauman 2019). While the methods vary, both prove helpful in oceanographic research and ocean conservation (e.g., Harcourt et al. 2019; Iverson et al. 2019; Williams et al. 2020).

5.7.1 Animal Tracking Tags

Acoustic tags, also known as pingers, transmit an acoustic signal that can be detected by a hydrophone. These tags, attached directly to the animal, require researchers to use a handheld receiver and follow the animal using a boat or to record the animal's movements using an array of moored hydrophones and data loggers that store the signals for later analysis. Acoustic tags offer subsurface information where signals from satellite-based global positioning systems (GPS) cannot transmit. California State Long Beach marine biologist Chris Lowe (*b.* 1963) and his "Shark Lab" use acoustic tags to track great white shark behavior along the Southern California coastline (e.g., Anderson et al. 2022). Such efforts help scientists understand shark behavior, an important consideration for public safety on beaches (Lowe 2019). Acoustic telemetry has also been integrated with AUVs to track sharks (e.g., Lowe et al. 2018).

A few nonacoustic tags deserve mention. Argos satellite tags appeared in the early 1980s when the newly established **Argos satellite system** (not to be confused with the Argo floats) came online as a tool for environmental monitoring. The Argos system provides a data uplink for tags placed on air-breathing animals, such as turtles and whales. Data are

FIGURE 5.49 UC Santa Cruz marine ecologist Ari Friedlaender attaches a tag to a blue whale off the coast of Southern California. The suction cup of the tag is harmless to the whale.

Credit: US Navy/Cascadia Research

FIGURE 5.50 Researchers release a tagged sea turtle off the coast of Saipan, the largest of the Northern Mariana islands.

Credit: NOAA CC BY 2.0

transmitted when the animals come to the surface. Alternatively, **pop-up satellite archival tags** (PSATs) permit researchers to study the behavior of organisms that do not typically spend time at the ocean surface, such as tuna, marlin, and halibut. The PSAT releases from the animal after a fixed time and floats to the surface, where it can transmit its data. **GPS tags** store data on the device until the animal is recaptured and the data downloaded from the tag. These devices are small enough to attach to coconut crabs (Krieger et al. 2012), stingrays (Martins et al. 2019), and marine birds (e.g., Peschko et al. 2020).

FIGURE 5.51 Tagged great white shark off Guadalupe Island, Baja California, Mexico. Note the tag at the base of its dorsal fin.

Credit: istock/RamonCarretero

5.7.2 Animal Oceanographers

Acoustic tags and electronic sensors have given rise to a new breed of oceanographer. The **animal oceanographers**, marine animals outfitted with electronic sensors, now routinely and boldly go where humans have rarely gone before. Animal oceanographers represent a new platform for observing the marine environment. After all, who knows the ocean better than the animals that inhabit it?

FIGURE 5.52
Animal oceanographers have become an indispensible tool for studying remote locations in the world ocean.

Credit: *Shaw and Chamberlin*

Animals can freely swim in places where ships and instruments have limited access. They also seek out areas not obvious to human observers or sensors, such as highly productive fronts and upwelling regions. Consequently, biotelemetry and biologging have become an integral part of the Global Ocean Observing System (McMahon et al. 2021). Animal oceanographers have provided hundreds of thousands of temperature and salinity profiles, especially in the Southern Ocean, an challenging environment for humans (e.g., Treasure et al. 2017).

FIGURE 5.53
Treated responsibly and ethically, animal oceanographers can provide essential data for understanding the ocean.

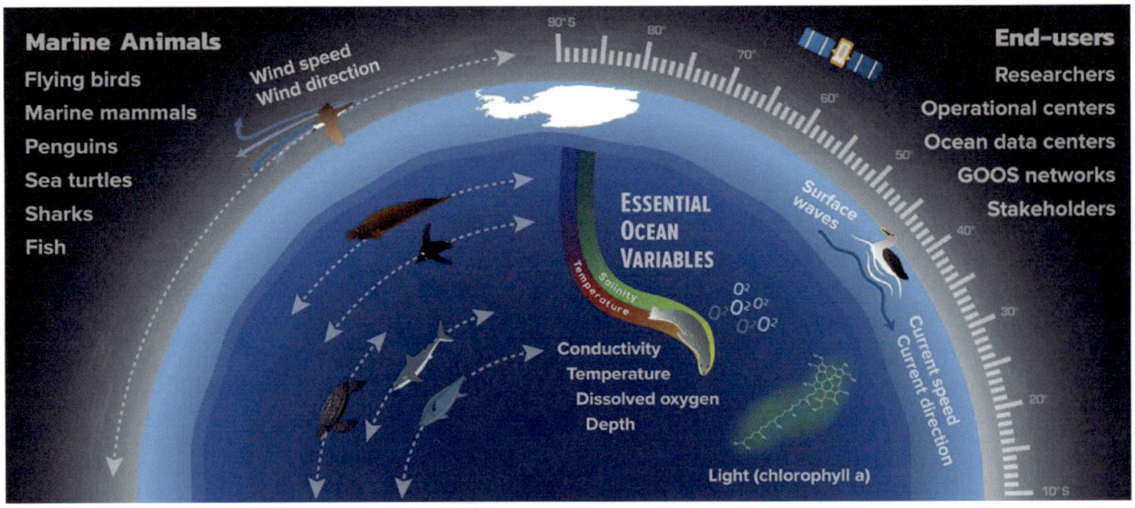

Credit: McMahon *et al.* (2021) CC BY 4.0

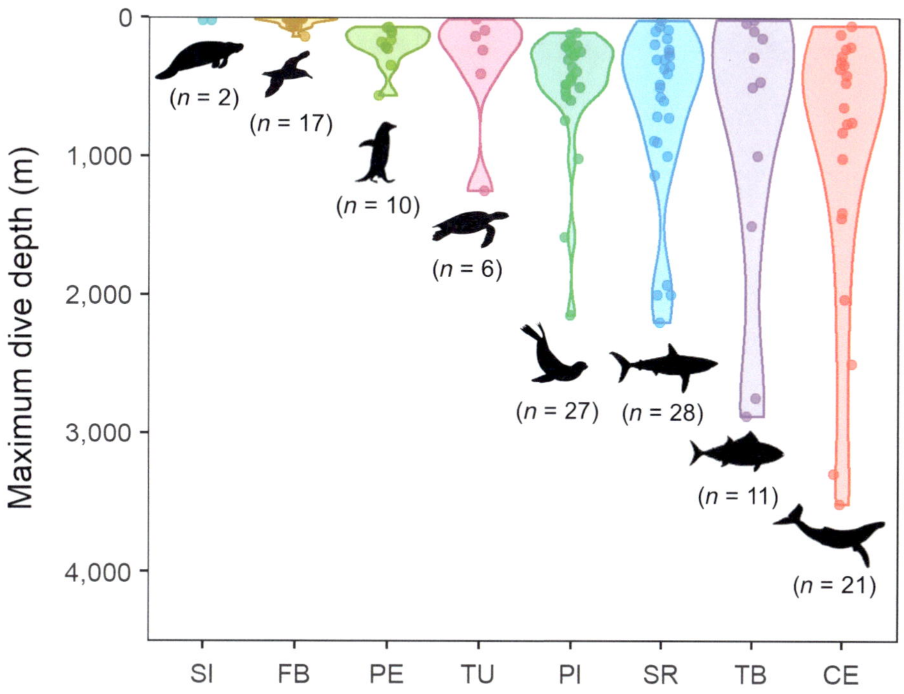

FIGURE 5.54 Maximum dive depths and number of species available as animal oceanographers.

Credit: March et al. (2020) CC BY 4.0

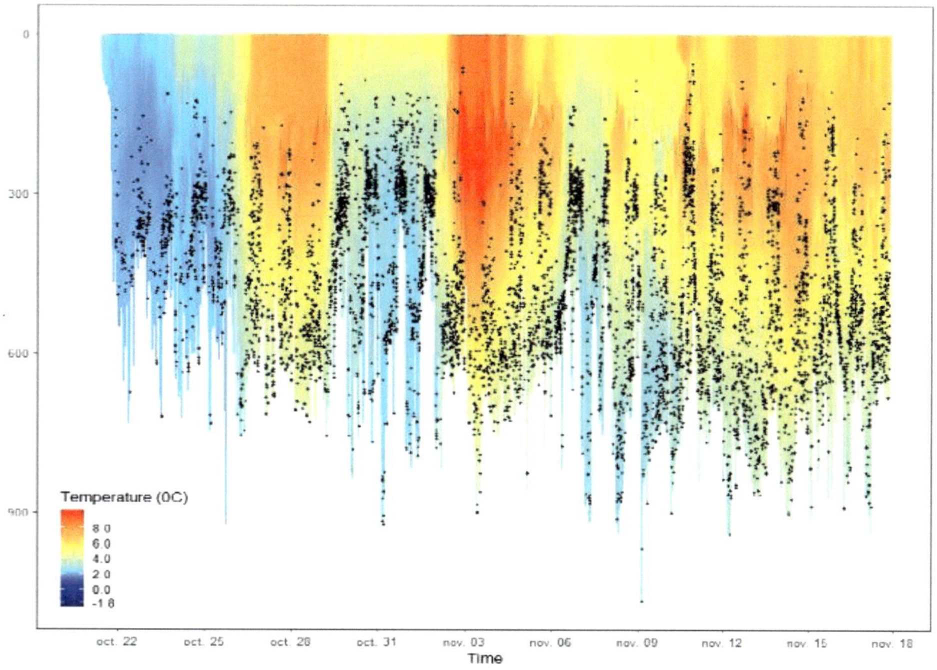

FIGURE 5.55 Data on temperature (colors) and acceleration (black dots) , indicative of feeding, obtained by a female southern elephant seal near the Kerguelan Islands in the Southern Ocean.

Credit: C. Guinet, Centre National de la Recherche Scientifique, McMahon et al. (2021) CC BY 4.0

5.8 Passive Acoustic Monitoring

Long used in antisubmarine warfare and research on fishes and marine mammals, hydrophones have reemerged as a twenty-first-century tool for observing the ocean (e.g., Au and Lammers 2016). Their modernization and deployment by geophysicists to record earthquakes and undersea eruptions have caused a resurgence of interest in them for bioacoustics research. These new devices, called **passive acoustic recorders**, combine modern electronics with digital recording devices to record sounds for long periods of time (Au and Lammers 2016). Automated recording has generated new insights into animal behavior, ecology, and conservation above and below water (Browning et al. 2017).

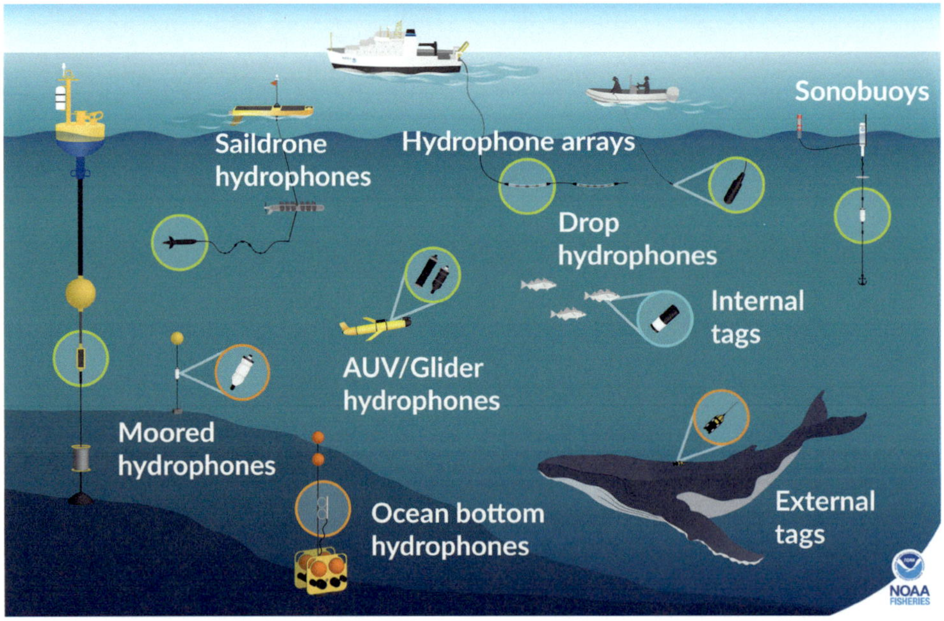

FIGURE 5.56 Passive acoustic technologies used by oceanographers.

Credit: NOAA Fisheries/Chamberlin

5.8.1 Ocean-Bottom Hydrophones

Modern passive acoustic recorders come in various shapes and sizes (Sousa-Lima et al. 2013). Because they emerged as a modification of ocean-bottom seismometers, they are called **ocean-bottom hydrophones**, listening devices temporarily or permanently moored on the seafloor.

Marine autonomous recording units developed by the Cornell Center for Conservation Bioacoustics were deployed in Massachusetts Bay in a 10-year study of Atlantic cod populations. This species produces sounds during spawning. The recordings provided valuable information on the seasonal behavior of the fishes and essential clues for their sustainable management (Caiger et al. 2020). Oceanographers working in the Gulf of Alaska used multiple ocean-bottom hydrophones developed at Scripps Institution of Oceanography—the high-frequency acoustic recording packages—to record and track the calls of fin whales, the second-largest whale in the ocean. The study revealed differences in the movements and behaviors of the whales. It demonstrated the usefulness of a multi-instrument approach for studying whale behavior (Wiggins and Hildebrand 2020).

FIGURE 5.57 Oceanographers deploy a hydrophone on a mooring to record ice shelf noises and marine mammals in the Ross Sea, Antarctica.

Credit: NOAA Pacific *Marine Environmental Laboratory*

5.8.2 Hydrophone on a Tag

Acoustic recorders can also be attached directly to animals. **Digital acoustic recording tags** (D-TAGs) incorporate hydrophones with other sensors to record an animal's vocalizations and movements. They also record environmental information, such as temperature and depth (e.g., Burgess et al. 1998; Johnson and Tyack 2003). D-TAG studies of southern right whales on calving grounds off Brazil detected low-volume grunts between lactating females and their calves that had not been observed previously using ocean-bottom hydrophones. The study also

FIGURE 5.58
A family of spotted dolphins in the Caribbean. Spotted dolphins use signature whistles to identify each other, the equivalent of human names. They use sound to hunt for prey beneath the sand and may even use it to disorient their prey. We have much to learn about animals' mastery of the underwater soundscape.

Dolphins live in a complex society with friends and relatives: they eat and hunt, raise young, share responsibilities, avoid predators, and resolve conflicts. They also live in a sensory world we can only imagine, full of different sounds, sights, and tastes, and a world we may never entirely understand. What we share with dolphins is their family lives, their daily challenges, and their fierce devotion to their offspring and community. It is here where the boundaries between humans and dolphins intersect and it is here were we must look for hope in their cultural preservation, as we hope for our own.

-- Dr. Denise Herzing, Dolphin Diaries (2011)
Wild Dolphin Project

Credit: istock/tswinner

FIGURE 5.59
A D-TAG affixed to a male beaked whale off Andros Island in the Bahamas.

Credit: US Navy

suggested that right whales may increase their "chatter" during social interactions (Dombroski et al. 2020). Perhaps one day scientists will develop the technology to allow us to chat with whales on smartphones.

5.8.3 Hydrophone on a Glider

Hydrophones attached to gliders offer an opportunity to simultaneously collect environmental and sound data. These data are crucial for understanding the spawning behavior of commercially important fishes. In 2015 scientists deployed a hydrophone equipped glider on the reefs off St. Croix in the Virgin Islands to study ecologically important snapper and groupers during their spawning cycle.

FIGURE 5.60
A hydrophone-equipped glider is ready for launch on a 20-day mission to listen to the reefs of St. Croix in the Virgin Islands.

Credit: NOAA National Ocean Service

5.9 Ocean Soundscapes

The 2010s witnessed a surge of interest in studies of **soundscapes**, the totality of sounds present in a particular region or habitat (e.g., Pijanowski et al. 2011). The field draws inspiration from studies of sounds in cities and natural environments (e.g., Southworth 1969; Schafer 1977). We now know that soundscapes reveal something about the behavior and social interactions of sound-makers, including us.

Credit: istock/chameleonseye

FIGURE 5.61 Young divers are often surprised to learn the ocean is filled with sounds.

The study of soundscapes belongs to a new field of science known as **soundscape ecology** (e.g., Pijanowski et al. 2011). Soundscape ecologists recognize three primary sources of sound: (1) **geophony**, sounds emanating from geologic and physical processes; (2) **biophony**, sounds originating from biological sources (except humans); and (3) **anthrophony**, sounds produced by human vocalizations and human activities.

Acoustic researchers have long known about **fish choruses**, sustained and elevated sound-making by fishes for an hour or more (e.g., Cato 1978; McWilliam et al. 2017). Fish choruses resemble the dawn chorus of

Credit: USGS (left); istock/Andrea Izzotti (middle); and istock/Yanlev (right)

FIGURE 5.62 (Left to right) Examples of geophony, biophony, and anthrophony in the ocean.

FIGURE 5.63

They may not sound like the Tabernacle Choir, but fish choruses represent an exciting area of research into the acoustic behavior of animals in the ocean environment.

Credit: Shaw and Chamberlin

birds or the peeps of frogs in the evening. The availability of passive acoustic recorders has provided new opportunities for studying ocean soundscapes (Lindseth and Lobel 2018). Fish choruses have been observed in nearly every ocean (e.g., McCauley 2012; Staaterman et al. 2014; Buscaino et al. 2016; Heenehan et al. 2019).

Studies of coral reef biophony have revealed sophisticated patterns in the timing of sounds on coral reefs. These studies support what has come to be known as **acoustic niches**, partitioning of the timing, intensity, or spectrum of sounds in a community of organisms to avoid overlap in sound-making (Krause 1993). It's the equivalent of everybody not talking at once. For example, observations using a passive acoustic recorder on a coral reef off Moorea (near Tahiti in French Polynesia) revealed distinct differences in sounds produced during the day versus the night. The study also revealed a sequence to the dominant sounds throughout

FIGURE 5.64

In recent years, marine biologists have discovered more than 1,000 vocalizing species, including these butterfly fish.

Credit: Ocean Image Bank/Jayne Jenkin

How well do you know your soundscape? Is it mostly natural or artificial? How does sound affect your mood and ability to focus? Sound reveals the rhythm of our planet and our lives.

the night, suggesting that animals "took turns" in their sound-making (Bertucci et al. 2020).

Ocean soundscapes exhibit complex and dynamic properties that provide invaluable insights into ocean ecosystems. They offer the potential for long-term monitoring of the biodiversity and conservation of ocean species (Lindseth and Lobel 2018). They also provide a much-needed baseline for establishing the impacts of human-generated sounds on marine organisms, the final topic in our chapter.

5.10 Ocean Noise

If you spend any time listening to the world around you—instead of blocking it out with earbuds—you will notice certain unpleasant sounds in your daily soundscape. Perhaps it's a leaf blower or nightly fireworks from your local magic kingdom. Maybe it's the din of traffic or an ambulance siren. Or maybe it's the hum of a refrigerator or the relentless tick of a clock at night. All these sounds define what scientists call **noise**, any unwanted or undesired sound in the environment. Of course, one person's "noise" may be another person's "music." Some people (like me) truly like the San Francisco punk band Flipper while others hate it (Fitzpatrick 2019). But in general, noise refers to sounds of a type and intensity that interfere with other acoustic sources or cause harm to them.

FIGURE 5.65
To some people it's music. To others, it's noise. You might say noise is in the ears of the beholder.

Credit: istock/IllustratedFritz

FIGURE 5.66

Sources of noise in the ocean include natural sounds, such as weather and earthquakes, animal sounds, especially marine mammals, and human-made sounds, such as ships, geophysical surveys, and even offshore wind farms.

***Credit:* NOAA Fisheries**

Numerous sources of ocean noise exist in the ocean: ships, offshore construction, sonar, seismic reflection studies, and more (e.g., Slabekoorn et al. 2010). Studies of ocean noise began during World War II to distinguish manmade sounds—especially enemy ships and submarines—from all other sounds (e.g., Knudsen et al. 1948). But as commercial shipping and recreational boating traffic grew, and science and industry adopted acoustic techniques for probing and observing the ocean, ocean noise also increased. Like the slowly boiling frog that doesn't notice the pot getting hotter, ocean noise grew steadily without attracting much attention. That quickly changed in the 1990s.

FIGURE 5.67

Scientists estimate that ship traffic accounts for the largest percentage of noise in the ocean.

***Credit:* istock/Crystal Calla**

5.10.1 The Ocean Acoustic Thermometry Experiment

I remember listening in rapt attention at the 1989 inaugural meeting of the Oceanography Society in Monterey Bay, California, as Scripps Institution of Oceanography physical oceanographer Walter Munk (1917–2019) described an idea to use low-frequency sound as a "thermometer" to observe ocean temperatures. The technique, known as **ocean acoustic tomography**, was first proposed by Munk and Massachusetts Institute of Technology physical oceanographer Carl Wunsch (Munk and Wunsch 1979). Because the speed of sound in the ocean increases as seawater temperatures increase, you can determine the degree to which the oceans are warming as long as you can accurately measure the sounds over long distances. By this time, scientific concern about human-caused global warming had grown considerably, and scientists were seeking independent ways to verify observations that Earth was warming. The idea seemed quite brilliant to me.

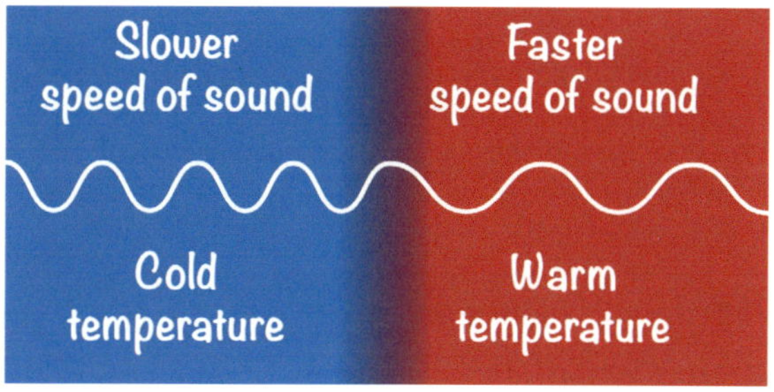

Credit: Chamberlin

FIGURE 5.68 The speed of sound across the global ocean can reveal how fast the ocean is warming.

Initial tests in January 1991 using an intense sound source at Heard Island, a remote location in the southern Indian Ocean, demonstrated that the sound signal could be detected throughout the world ocean. The stage was set for a 10-year experiment, the Acoustic Thermometry of Ocean Climate (ATOC) project

Credit: NASA

FIGURE 5.69 The glaciated peak of Mount Mawson on Heard Island, one of the most remote islands in the world.

FIGURE 5.70
The ATOC feasibility test sought to determine whether a sound source at Heard Island could be detected at global listening stations (starred circles). Concern for marine mammals halted the project.

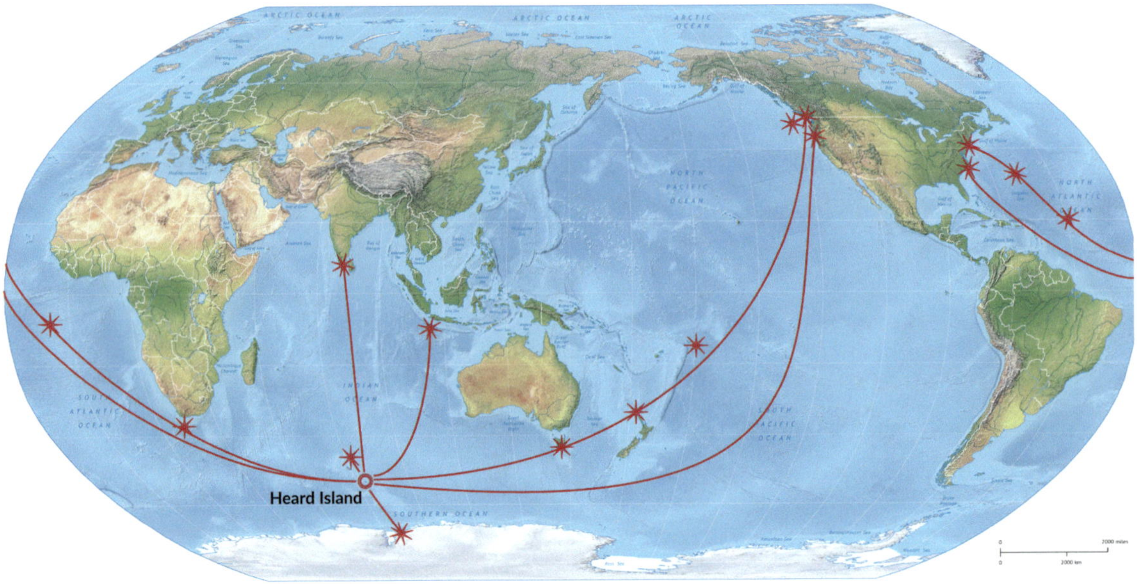

Credit: Chamberlin after Baggeroer and Munk (1992)

(Baggeroer and Munk 1992). Except the global experiment never happened.

An article in *Science* in April 1990 on ATOC caught the eye of marine mammal biologists (Cohen 1991), and a March 1994 syndicated LA Times article raised a public outcry. Scientists and the public began to voice concerns about the effects of the high-intensity sound on marine mammals. Because the animals were protected under the Marine Mammal Act of 1972, the National Research Council (NRC) responded by commissioning a study to review what is known about the effects of ocean noise and low-frequency sound on marine mammals (NRC 1994). The studies were inconclusive (NRC 2000, 2003), but the public backlash proved too damaging (e.g., Potter 1994). Regional deployments demonstrated the usefulness of acoustic

FIGURE 5.71
Concerns about impacts of high decibel noises on marine mammal health eventually halted a global-scale ATOC experiment.

Credit: NOAA CC BY 2.0

FIGURE 5.72 Conflicting opinions over complex problems often pit well-intentioned people against each other.

Credit: istock/Cemile Bingol, modified by Chamberlin

tomography for observing ocean temperatures (e.g., Dushaw 2014). Still, the complete project as envisioned by Munk never came to pass.

The controversy over the ATOC experiment caught scientists by surprise. It also pitted environmentalists against environmentalists in a battle of concerns between global warming and marine mammal protection. It's a cautionary tale for scientists to learn to better communicate with the public and a call for science educators to help the public better understand the nature of science. As Potter (1994) wrote nearly three decades ago: "The combination of increasing public concern with environmental issues, disenchantment over scientific and governmental integrity, and the widely accessible information highway combined to produce an explosive mixture which government, the public, and scientists need to learn how to handle more responsibly in the future."

The story has a happy ending, however. In 2020, California Institute of Technology seismologist Wenbo Wu and colleagues published a modified method for ocean acoustic tomography using naturally occurring seismic waves, what they called seismic ocean thermometry (Wu et al. 2020). Comparing the speeds of ocean-traveling seismic waves across the

FIGURE 5.73 Seafloor earthquakes off Sumatra, near this location on Ko Lanta Island in Thailand, provided a seismic source for establishing ocean seismic thermometery.

Credit: istock/petdcat

equatorial Indian Ocean from 2005 to 2016, they observed temperature trends that compared favorably with those measured by Argo floats and estimated using computer models of ocean temperature. As they express at the end of their paper, "Global seismic ocean thermometry can thus substantially enhance our capability to monitor ocean warming." While their methods and observations will require further confirmation by the scientific community, the promise of acoustic tomography—as first envisioned more than four decades ago—may finally be realized (e.g., Dushaw 2022).

5.10.2 Ocean Noise in the Twenty-First Century

Concerns over ATOC have also focused public attention on other acoustic technologies and ocean noise in general (NRC 2003). Seismic reflection studies using airguns drew the attention of marine mammal scientists as a possible cause for beaked whale strandings in the Gulf of California (Malakoff 2002). A federal judge ordered a halt to the work (LA Times 2002). That prompted an ocean conservation group to conduct and publish a study on the effects of seismic airgun testing on marine mammals with the provocative title, A *Deaf Whale Is a Dead Whale*

FIGURE 5.74 Noise levels have increased in some parts of the world ocean, but other locations appear quieter.

Credit: Shaw and Chamberlin

(Heulsenbeck and Wood 2013). The report and headline made newspapers (e.g., USA Today 2013). They often resurface when stories about seismic reflection studies appear (e.g., New Internationalist 2019). Are such reports overly dramatic, focusing only on the negative issues to stir public emotion?

In a 2011 article in *Oceanography*, Boyd et al. ask, "Does the noise made by humans harm marine life?" As with many scientific questions, the devil is in the details. How do you characterize harmful ocean noise? How do you define harm? What types of marine life should be considered? These are not easy questions to answer.

As researchers in this field emphasize, a great deal of additional study is needed (Tyack et al. 2015; Williams et al. 2015; Erbe et al. 2019; deQuiros et al. 2019; Duarte et al. 2021). Human-generated noise can be defined in terms of its acoustic characteristics and variability in time and space (e.g., Lindgren and Wilewska-Bien 2016). Unfortunately, measurements of these variables are lacking (Erbe et al. 2019). Collectively, ocean ship noise along the US West Coast has increased since the 1960s. Current levels appear steady or decreasing. In other regions of the ocean, such as the

FIGURE 5.75 Strandings and deaths of beaked whales have been linked to sonar and seismic reflection air guns. The devices may cause the deep-diving animals to surface too quickly and suffer embolisms.

***Credit:* US Navy**

Indian Ocean, ship traffic has increased dramatically (Andrew et al. 2011; Tournade 2014; Miksis-Olds and Nichols 2016; Ragland et al. 2022).

The lack of long-term comprehensive data on ocean noise in US waters prompted NOAA and the National Parks Service to establish in 2014 the Ocean Noise Reference Station Network, described as "an array of . . . calibrated autonomous passive acoustic recorders" (Haver et al. 2018). Efforts to address the global data gap have begun under the guidance of the International Quiet Ocean Experiment, a "program of research, observation, and modeling to better characterize ocean sound fields and to promote understanding of the effects of sound on marine life" (Tyack et al. 2015; Duarte et al. 2021).

FIGURE 5.76 Naval vessel sonar has been implicated in the strandings of marine mammals (e.g., Moretti et al. 2014).

***Credit:* US Navy**

FIGURE 5.77 Oceanographers deployed a hydrophone in the Challenger Deep to record sounds in the deepest part of the ocean. Better monitoring will improve our ability to understand spatial and temporal patterns of change in ocean noise and their potential impacts.

Credit: US Coast Guard/Tara Molle

Observing and characterizing harm to marine animals proves equally challenging. While there is evidence that military sonar may lead to beach strandings and deaths for some species, the evidence is not direct, and these occurrences are rare. Loss of hearing—temporarily or permanently—due to loud and sustained sound has been shown in captive animals. How these results apply to animals in the wild, who can move freely and generally avoid harmful sounds, is unknown. Effects on animal behavior, health, survival, and reproductive rates remain unclear (e.g. Weilgart 2017).

Nevertheless, new clues have emerged on the impacts of ocean noise on fishes and mammals. The behavior of Atlantic cod appears disrupted by seismic reflection surveys (van der Knapp et al. 2021), and Pacific salmon and herring alter their behavior in response to boat noise (van der Knapp et al. 2022). A study of vocalizations by humpback whales revealed that the whales sing louder when ocean noise levels are louder (e.g., Dunlop et al. 2014; Guazzo et al. 2020). This behavior of increasing sound level in response to increased background noise is known as the Lombard effect, named after French doctor Étienne Lombard (1869–1920). He, like you, noted that humans talk louder when their environment is louder (e.g., Zollinger and Brumm 2011). Other marine mammals, including minke whales and bottlenose dolphins, also exhibit the Lombard effect in response to increasing ocean noise (Kragh et al. 2019; Helbe et al. 2020;).

As unsatisfactory as the answer "we don't know yet" can be to a public accustomed to definitive answers, we can gain solace in the knowledge that research in this field is proliferating (Williams et al. 2015). While scientists acknowledge that "we are unlikely to resolve this challenge quickly or completely" (Tyack et al. 2015), the availability of new tools and the establishment of long-term acoustic monitoring networks provide promising avenues for a better understanding of ocean noise and its impacts. Public outreach and education—and transparent communication with public, commercial, and governmental stakeholders—will help establish the societal importance and public trust for the continued support of ocean acoustic science (Spence et al. 2022).

FIGURE 5.78 Young male humpback whales-perhaps sharing a song-off Moorea in French Polynesia.

Credit: Ocean Image Bank/Ron Watkins

5.11 Chapter References

Adams, K. T. 1942. *Hydrographic Manual*, Special Publication 143 (Revised 1942 Edition). Washington DC: US Government Printing Office. https://doi.org/10.5962/bhl.title.16377

Anderson, James M. Emily Spurgeon, Brian S. Stirling, Jack May III, Patrick T. Rex, Bobby Hyla, Steve McCullough, Marten Thompson, and Christopher G. Lowe. 2022. "High Resolution Acoustic Telemetry Reveals Swim Speeds and Inferred Field Metabolic Rates in Juvenile White Sharks (*Carcharodon carcharias*)." *PLOS One* 17(6): e0268914. https://doi.org/10.1371/journal.pone.0268914

Andrew, Rex K., Bruce M. Howe, and James A. Mercer. 2011. "Long-Time Trends in Ship Traffic Noise for Four Sites off the North American West Coast" *The Journal of the Acoustical Society of America* 129(2): 642–651. https://doi.org/10.1121/1.3518770

Au, Whitlow W. L., and Marc O. Lammers. 2016. "Introduction: Listening in the Ocean." In *Listening in the Ocean: New Discoveries and Insights on Marine Life from Autonomous Passive Acoustic Recorders*. Edited by Whitlow W. L. Au and Marc O. Lammers. New York: Springer. https://doi.org/10.1007/978-1-4939-3176-7_1

Baggeroer, Arthur, and Walter Munk. "The Heard Island Feasibility Test." *Physics Today* 45(9): 22–30. https://doi.org/10.1063/1.881317

Bauman, Karen. 2019. "Wildlife Technologies." In *Fowler's Zoo and Wild Animal Medicine: Current Therapy*, Volume 9. Edited by R. Eric Miller, Nadine Lamberski, and Paul P. Calle, 11-15. St. Louis: Elsevier. https://doi.org/10.1016/B978-0-323-55228-8.00003-5

Bertucci, Frédéric, Katy Maratrat, Cécile Berthe, Marc Besson, Ana Sofia Guerra, Xavier Raick, Frank Lerouvreur, David Lecchini, and Eric Parmientier. 2020. "Local Sonic Activity Reveals Potential Partitioning in a Coral Reef Fish Community." *Oecologia* 193: 125–134. https://doi.org/10.1007/s00442-020-04647-3

Biescas, B., V. Sallarès, J. L. Pelegri, F. Machin, R. Carbonell, G. Buffett, J. J. Dañobeitia, and A. Calahorrano. 2008. "Imaging Meddy Fine Structure Using Multichannel Seismic Reflection Data." *Geophysical Research Letters* 35: L11609. https://doi.org/10.1029/2008GL033971

Block, Barbara A., Christopher M. Holbrook, Samantha E. Simmons, Kim N. Holland, Jerald, S. Ault, Daniel P. Costa, Bruce R, Mate, Andrew C. Seitz, Michael D. Arendt, John C. Payne, Behzad Mahmoudi, Peter Moore, James M. Price, J. Jacob Levenson, Doug Wilson, and Randall E. Kochevar. 2016. "Toward a National Animal Telemetry Network for Aquatic Observations in the United States." *Animal Biotelemetry* 4: 6. https://doi.org/10.1186/s40317-015-0092-1

Blondel, Philippe. 2009. *Handbook of Side-Scan Sonar*. Heidelberg: Springer Berlin. https://doi.org/10.1007/978-3-540-49886-5

Bongiovanni, Cassandra, Heather A. Stewart, and Alan J. Jamieson. 2021. "High-Resolution Multibeam Sonar Bathymetry of the Deepest Place in Each Ocean." *Geoscience Data Journal* 9(1): 1–16. https://doi.org/10.1002/gdj3.122

Boyd, Ian L., George Frisk, Ed Urban, Peter Tyack, Jesse Ausubel, Sophie Seeyave, Doug Cato, Brandon Southall, Michael Weise, Rex Andrew, Tomonari Akamatsu, René Dekeling, Christine Erbe, David Farmer, Roger Gentry, Tom Gross, Anthony Hawkins, Fenghua Li, Kathy Metcalf, James H. Miller, David Moretti, Christian Rodrigo, and Tomio Shinke. 2015. "An International Quiet Ocean Experiment." *Oceanography* 24(2): 174–181. https://doi.org/10.5670/oceanog.2011.37

Brinkløv, Signe, M. Brock Fenton, and John M Ratcliffe. 2013. "Echolocation in Oilbirds and Swiftlets." *Frontiers in Physiology* 4: 123. https://doi.org/10.3389/fphys.2013.00123

Browning, Ella, Rory Gibb, Paul Glover-Kapfer, and Kate E. Jones. 2017. *Passive Acoustic Monitoring in Ecology and Conservation*. WWF Conservation Technology Series 1(2). Woking: WWF-UK. https://www.wwf.org.uk/sites/default/files/2019-04/Acousticmonitoring-WWF-guidelines.pdf

Buffet, G. G., R. Carbonell, and Jaume Almera. 2011. "Seismic Oceanography as a Tool to Monitor Climate Change." *The Eggs* 35: 5–8. https://www.egu.eu/newsletter/?offset=12

Burgess, W. C., P. L. Tyack, B. J. Le Boeuf, and D. P. Costa. 1998. "A Programmable Acoustic Recording Tag and First Results from Free-Ranging Northern Elephant Seals." *Deep Sea Research Part II: Topical Studies in Oceanography* 45(7): 1327–1351. https://doi.org/10.1016/S0967-0645(98)00032-0

Buscaino, Giuseppa, Maria Ceraulo, Nadia Pieretti, Valentina Corrias, Almo Farina, Francesco Filiciotto, Vincenzo Maccarone, Rosario Grammauta, Francesco Caruso, Alonge Giuseppe, and Salvatore Mazzola. 2016. "Temporal Patterns in the Soundscape of the Shallow Waters of a Mediterranean Marine Protected Area." *Scientific Reports* 6: 34230. https://doi.org/10.1038/srep34230

Caiger, Paul E., Micah J. Dean, Annamaria I. DeAngelis, Leila T. Hatch, Aaron N. Rice, Jenni A. Stanley, Chris Thilke, Douglas R. Zemeckis, and Sofie M. Van Parijs. 2020. "A Decade of Monitoring Atlantic Cod *Gadus morhua* Spawning Aggregations in Massachusetts Bay Using Passive Acoustics." *Marine Ecology Progress Series* 635: 89–103. https://doi.org/10.3354/meps13219

Cato, Douglas H. 1978. "Marine Biological Choruses Observed in Tropical Waters near Australia." *The Journal of the Acoustical Society of America* 64(3): 736–743. https://doi.org/10.1121/1.382038

Cisewski, Boris, Hjálmar Hátún, Inga Kristiansen, Bogi Hansen, Karin Margretha H. Larsen, Sólvá Káradóttir Eliasen, and Jan Arge Jacobsen. 2021. "Vertical Migration of Pelagic and Mesopelagic Scatterers From ADCP Backscatter Data in the Southern Norwegian Sea." *Frontiers in Marine Science* 7: 542386. https://doi.org/10.3389/fmars.2020.542386

Cohen, Jon. 1991. "Was Underwater "Shot" Harmful to the Whales?" *Science* 252(5008): 912–914. https://doi.org/10.1126/science.252.5008.912

Crutchley, Gareth J., and Heidrum Kopp. 2018. "Reflection and Refraction Seismic Methods." In *Submarine Geomorphology*. Edited by Aaron Micallef, Sebastian Krastel, and Alessandra Savini. 43–62. Cham: Springer Geology. https://doi.org/10.1007/978-3-319-57852-1_4

D'Amico, Angela, and Richard Pittenger. 2010. A Brief History of Active Sonar. *Aquatic Mammals* 35(4): 426–434. https://doi.org/10.1578/AM.35.4.2009.426

Dean, Dennis. "Robert Mallet and the Founding of Seismology." *Annals of Science* 48(1): 39–67. https://doi.org/10.1080/00033799100200121

de Quirøs, Y. Bernaldo, A. Fernandez, R. W. Baird, R. L. Brownell, N. Aguilar de Soto, D. Allen, M. Arbelo, M. Arregui, A. Costidis, A. Fahlman, A. Frantzis, F. M. D. Gulland, M. Iñíguez, M. Johnson, A. Komnenou, H. Koopman, D. A. Pabst, W. D. Roe, E. Sierra, M. Tejedor, and G. Schorr. 2019. "Advances in Research on the Impacts of Anti-submarine Sonar on Beaked Whales." *Proceedings of the Royal Society B* 286(1895): 20182533. https://doi.org/10.1098/rspb.2018.2533

DeRobertis, Alex, Robert Levine, and Christopher D. Wilson. 2017. "Can a Bottom-Moored Echo Sounder Array Provide a Survey-Comparable Index of Abundance?" *Canadian Journal of Fisheries and Aquatic Sciences* 75(4): 629–640. https://doi.org/10.1139/cjfas-2017-0013

Dombrowski, Julia R. G., Susan E. Parks, Paulo A. C. Flores, Lucía M. Martín López, K. Alex Shorter, and Karina R. Groch. 2020. "Animal-Borne Tags Provide Insights into the Acoustic Communication of Southern Right Whales (*Eubalaena australis*) on the Calving Grounds." *The Journal of the Acoustical Society of America* 147: EL498. https://doi.org/10.1121/10.0001391

Duarte, Carlos M., Lucille Chapuis, Shaun P. Collin, Daniel P. Costa, Reny P. Devassy, Victor M. Eguiluz, Christine Erbe, Timothy A. C. Gordon, Benjamin Halpern, Harry R. Harding, Michelle N. Havlik, Mark Meekan, Nathan D. Merchant, Jennifer L. Miskis-Olds, Miles Parson, Milica Predragovic, Andrew N. Radford, Craig A. Radford, Stephen D. Simpson, Hans Slabbekoorn, Erica Staaterman, Ilse C. Van Opzeeland, Jana Winderen, Xiangliang Zhang, and Frances Juanes. 2021. "The Soundscape of the Anthropocene Ocean." *Science* 371(6529): eaba4658. https://doi.org/10.1126/science.aba4658

Dunlop, Katherine M., Douglas H. Cato, and Michael J. Noad. 2014. "Evidence of a Lombard Response in Migrating Humpback Whales (*Megaptera novaeangliae*)." *The Journal of the Acoustical Society of America* 136: 430–437. https://doi.org/10.1121/1.4883598

Dunlop, Katherine M., Toby Jarvis, Kelly J. Benoit-Bird, Chad M. Waluk, David W. Caress, Hans Thomas, and Kenneth L. Smith, Jr. 2018. "Detection and Characterisation of Deep-Sea Benthopelagic Animals from an Autonomous Underwater Vehicle with a Multibeam Echosounder: A Proof of Concept and Description of Data-Processing Methods" *Deep Sea Research Part I: Oceanographic Research Papers* 134: 64–79. https://doi.org/10.1016/j.dsr.2018.01.006

Dunlop, K. M., K. J. Benoit-Bird, C. M. Waluk, R. G. Henthorn. 2020. "Ecological Insights into Abyssal Bentho-Pelagic Fish at 4000 m Depth Using a Multibeam Echosounder on a Remotely Operated Vehicle." *Deep Sea Research Part II: Topical Studies in Oceanography* 173: 104679. https://doi.org/10.1016/j.dsr2.2019.104679

Dushaw, Brian. 2015. "Acoustic Tomography, Ocean." In *Encyclopedia of Remote Sensing*. Edited by E. G. Njoku. Encyclopedia of Earth Science Series. New York: Springer. https://doi.org/10.1007/978-0-387-36699-9_211

Dunshaw, Brad D. 2022. "Surprises in Physical Oceanography: Contributions from Ocean Acoustic Tomography." *Tellus A: Dynamic Meteorology and Oceanography* 74: 33–67. https://doi.org/10.16993/tellusa.39

Earth Institute. 2020. "Marie Tharp's Adventures in Mapping the Seafloor, In Her Own Words." Earth Institute. July 24, 2020. https://news.climate.columbia.edu/2020/07/24/marie-tharp-connecting-dots/

Erbe, Christine, Sarah A. Marley, Renée P. Schoeman, Joshua N. Smith, Leah E. Trigg, and Clare Beth Embling. 2019. "The Effects of Ship Noise on Marine Mammals—A Review." *Frontiers in Marine Science* 6: 606. https://doi.org/10.3389/fmars.2019.00606

Fitzpatrick, Will. 2019. "Life Is The Only Thing Worth Living For: Flipper Interviewed." Clash. July 7, 2019. https://www.clashmusic.com/features/life-is-the-only-thing-worth-living-for-flipper-interviewed

Fornshell, John A., and Allesandra Tesei. 2013. "The Development of SONAR as a Tool in Marine Biological Research in the Twentieth Century." *International Journal of Oceanography* 2013: 678621. https://doi.org/10.1155/2013/678621

Grothues, Thomas M., Arthur E. Newhall, James F. Lynch, Kaela S. Vogel, and Glen G. Gawarkiewicz. 2017. "High-Frequency Side-Scan Sonar Fish Reconnaissance by Autonomous Underwater Vehicles." *Canadian Journal of Fisheries and Aquatic Sciences* 74(2): 240–255. https://doi.org/10.1139/cjfas-2015-0301

Guazzo, Regina A., Tyler Helble, Gabriela C Alongi, Ian N. Durbach, Cameron R. Martin, Stephen W. Martin, and E. Elizabeth Henderson. 2020. "The Lombard Effect in Singing Humpback Whales: Source Levels Increase as Ambient Ocean Noise Levels Increase." *The Journal of the Acoustical Society of America* 148: 542–555. https://doi.org/10.1121/10.0001669

Gunn, Kathryn L., Nicky White, and Colm-cille P. Caulfield. 2020. "Time-Lapse Seismic Imaging of Oceanic Fronts and Transient Lenses Within South Atlantic Ocean." *Journal of Geophysical Research* 125(7): e2020JC016293. https://doi.org/10.1029/2020JC016293

Guzman, Hector M. and Richard Condit. 2017. "Abundance of Manatees in Panama Estimated from Side-Scan Sonar." *Wildlife Society Bulletin* 41(3): 556–565. https://doi.org/10.1002/wsb.793

Hansen, Roy E. 2019. "Mapping the Ocean Floor in Extreme Resolution Using Interferometric Synthetic Aperture Sonar." *Proceedings of Meetings on Acoustics* 38(1): 055003. https://doi.org/10.1121/2.0001058

Harcourt, Rob, Ana M. M. Sequeira, Xuelei Zhang, Fabien Roquet, Kosei Kamatsu, Michelle Heupel, Clive McMahon, Fred Whoriskey, Mark Meekan, Gemma Carroll, Stephanie Brodie, Colin Simpfendorfer, Mark Hindell, Ian Johnson, Daniel P. Costa, Barbara Block, Mônica Muelbert, Bill Woodward, Mike Weise, Jim Aarestrup, Martin Biuw, Lars Boehme, Steven J. Bograd, Dorian Cazau, Jean-Benoit Charrassin, Steven J. Cooke, Paul Cowley, P. J. Nico de Bruyn, Tiphaine Jeanniard du Dot, Carlos Duarte, Víctor M. Eguíluz, Luciana C. Ferriera, Juan Fernández-Gracia, Kimberly Goetz, Yusuke Goto, Christophe Guinet, Mike Hammill, Graeme C. Hays, Elliott L. Hazen, Luis A. Hückstadt, Charlie Huveneers, Sara Iverson, Saifullah Arifin Jaaman, Kongkiat Kittiwattanawong, Kit M. Kovacs, Christian Lydersen, Tim Moltmann, Masaru Naruoka, Lachlan Phillips, Baptiste Picard, Nuno Queiroz, Gilles Reverdin, Katsufumi Sato, David W. Sims, Eva B. Thorstad, Michele Thums, Anne M. Treasure, Andrew W. Trites, Guy D. Williams, Yoshinari Yonehara, and Mike A. Fedak. 2019. "Animal-Borne Telemetry: An Integral Component of the Ocean Observing Toolkit." *Frontiers in Marine Science* 6: 326. https://doi.org/10.3389/fmars.2019.00326

Haver, Samara M., Jason Gedamke, Leila T. Hatch, Robert P. Dziak, Sofie Van Parijs, Megan F. McKenna, Jay Barlow, Catherine Berchok, Eva DiDonato, Brad Hanson, Joseph Haxel, Marla Holt, Danielle Lipski, Haru Matsumoto, Christian Meinig, David K. Mellinger, Sue E. Moore, Erin M. Oleson, Melissa S. Soldevilla, and Holger Klinck. 2018. "Monitoring Long-Term Soundscape Trends in US Waters: The NOAA/NPS Ocean Noise Reference Station Network." *Marine Policy* 90: 6–13. https://doi.org/10.1016/j.marpol.2018.01.023

Hawley, J. H. 1931. *Hydrographic Manual*, Special Publication 143. Washington DC: US Government Printing Office. https://www.blueplanetpublishing.com/wp-content/uploads/2023/10/HYDROGRAPHIC-MANUAL-143_hawley_1931.pdf

Hedström, H., and R. Kollert. 1949. "Seismic Sounding of Shallow Depths." *Tellus* 1(4): 24–36. https://doi.org/10.3402/tellusa.v1i4.8519

Heezen, Bruce C., Marie Tharp, and Maurice Ewing. 1959. *The Floors of the Oceans: 1. The North Atlantic*. Volume 65. New York: The Geological Society of America. https://doi.org/10.1130/SPE65

Heenehan, Heather, Joy E. Stanistreet, Peter J. Corkeron, Laurent Bouveret, Julien Chalifour, Genevieve E. Davis, Angiolina Henriquez, Jeremy J. Kiszka, Logan Kline, Coarline Reed, Omar Shamir-Reynoso, Fabien Védie, Wijnand De Wolf, Paul Hoetjes, and Sofie M. Van Parijs. 2019. "Caribbean Sea Soundscapes: Monitoring Humpback Whales, Biological Sounds, Geological Events, and Anthropogenic Impacts of Vessel Noise." *Frontiers in Marine Research* 6: 347. https://doi.org/10.3389/fmars.2019.00347

Helbe, Tyler A., Regina A. Guazzo, Cameron R. Martin, Ian N. Durbach, Gabriela C. Alongi, Stephen W. Martin, John K. Boyle, and E. Elizabeth Henderson. 2020. "Lombard Effect: Minke Whale Boing Call Source Levels Vary with Natural Variations in Ocean Noise." *The Journal of the Acoustical Society of America* 147(2): 698–712. https://doi.org/10.1121/10.0000596

Hersey, J. B. 1977. "A Chronicle of Man's Use of Ocean Acoustics." *Oceanus* 20(2): 8–21. https://www.biodiversitylibrary.org/page/1514044

Herzing, Denise L. 2011. *Dolphin Diaries: My 25 Years with Spooted Dolphins in the Bahamas*. New York: St. Martin's Press. https://us.macmillan.com/books/9781429987448/dolphindiaries

Holbrook, Steven, Pedro Páramo, Scott Pearse, and Raymond W. Schmitt. 2003. "Thermohaline Fine Structure in an Oceanographic Front from Seismic Reflection Profiling." *Science* 301(5634): 821–824. https://doi.org/10.1126/science.1085116

Holbrook, W. Steven, Ilker Fer, Raymond W. Schmitt, Daniel Lizarralde, Jody M. Klymak, L. Cody Helfrich, and Robert Kubichek. 2013. "Estimating Oceanic Turbulence Dissipation from Seismic Images." *Journal of Atmospheric and Oceanic Technology* 30(8): 1767–1788. https://doi.org/10.1175/JTECH-D-12-00140.1

Holliday, D. V., and R. E. Pieper. 1995. "Bioacoustical Oceanography at High Frequencies." ICES *Journal of Marine Science* 52(3–4): 279–296. https://doi.org/10.1016/1054-3139(95)80044-1

Holman, Zoe. 2019. "Greece Opens Its Arms—and Seas—to Hydrocarbon Giants." New Internationalist. August 2, 2019. https://newint.org/features/2019/08/02/greece-opens-its-arms-and-seas-hydrocarbon-giants

Howe, Bruce M., Jennifer Miksis-Olds, Eric Rehm, Hanne Sagen, Peter F. Worcester, and Georgios Haralabus. 2019. "Observing the Oceans Acoustically." *Frontiers in Marine Science* 6: 426. https://doi.org/10.3389/fmars.2019.00426

Huelsenbeck, Matthew and Caroline Wood. 2013. "A Deaf Whale is A Dead Whale: Seismic Airgun Testing for Oil and Gas Threatens Marine Life and Coastal Economies." Oceana. April 2013. https://oceana.org/press-releases/marine-life-and-coastal-economies-threatened-seismic-airgun-use-atlantic/

Iverson, Sara J., Aaron T. Fisk, Scott G. Hinch, Joanna Mills Flemming, Steven J. Cooke, and Frederick G. Whoriskey. 2019. "The Ocean Tracking Network: Advancing Frontiers in Aquatic Science and Management." *Canadian Journal of Fisheries and Aquatic Sciences* 76(7): 1041–1051. https://doi.org/10.1139/cjfas-2018-0481

Johnson, Mark P., and Peter L. Tyack. 2003. "A Digital Acoustic Recording Tag for Measuring the Response of Wild Marine Mammals to Sound." *IEEE Journal of Oceanic Engineering* 28(1): 3–12. https://doi.org/10.1109/JOE.2002.808212

Johnson, Martin. 1948. "Sound as a Tool in Marine Ecology, from Data on Biological Noises and the Deep Scattering Layer." *Journal of Marine Research* 7(3): 443–458. https://peabody.yale.edu/explore/publications/journal-marine-research

Kenny, A. J., I Cato, M. Desprez, G. Fader, R. T. E. Schüttenhelm, and J. Side. 2003. "An Overview of Seabed-Mapping Technologies in the Context of Marine Habitat Classification." *ICES Journal of Marine Science* 60(2): 411–418. https://doi.org/10.1016/S1054-3139(03)00006-7

Klien, Elias. 1968. "Underwater Sound and Naval Acoustical Research and Applications before 1939." *Acoustical Society of America* 43(5): 931–947. https://doi.org/10.1121/1.1910963

Knudsen, Vern O., R. S. Alford, and J. W. Emling. 1948. "Underwater Ambient Noise." *Journal of Marine Research* 7(3): 410–429. https://peabody.yale.edu/explore/publications/journal-marine-research

Kragh, Ida M., Katherine McHugh, Randall S. Well, Laela S. Sayigh, Vincent M. Janik, Peter L. Tyack, and Frants H. Jensen. 2019. "Signal-Specific Amplitude Adjustment to Noise in Common Bottlenose Dolphins (*Tursiops truncatus*)." *Journal of Experimental Biology* 222(23): jeb216606. https://doi.org/10.1242/jeb.216606

Krause, Bernie. 1993. "The Niche Hypothesis: A Virtual Symphony of Animal Sounds, the Origins of Musical Expression and the Health of Habitats." *The Soundscape Newsletter* 6: 4–6. June 1993. https://www.sfu.ca/sonic-studio-webdav/WSP_Doc/Newsletters/Number6.pdf

Krieger, Jakob, Ronald Grandy, Michelle M. Drew, Susanne Erland, Marcus C. Stensmyr, Steffen Harzsch, and Bill S. Hansson. 2012. "Giant Robber Crabs Monitored from Space: GPS-Based Telemetric Studies on Christmas Island (Indian Ocean)." *PLOS One* 7(11): e49808. https://doi.org/10.1371/journal.pone.0049809

Lindgren, J. Frederick, and Magda Wilewska-Bien. 2016. "Anthropogenic Noise." In *Shipping and the Environment: Improving Environmental Performance in Marine Transportation*. Edited by Karin Andersson, Selma Brynolf, J. Fredrik Lindgren, and Magda Wilewska-Bien. 229–235. Berlin: Springer-Verlag. https://doi.org/10.1007/978-3-662-49045-7_6

Lindseth, Adelaide V., and Phillip S. Lobel. 2018. "Underwater Soundscape Monitoring and Fish Bioacoustics: A Review." *Fishes* 3(3): 36. https://doi.org/10.3390/fishes3030036

Lowe, Christopher G. 2019. "California's Great Whites: What's going on?" Lecture, Fullerton Museum Center, Fullerton, October 5, 2019. https://m.facebook.com/events/fullerton-museum-center/californias-great-whites-whats-going-on/774810709614614/?locale=sw_KE&_rdr

Lowe, Christopher G., Connor F. White, and Christopher M. Clark. 2018. "Use of Autonomous Vehicles for Tracking and Surveying of Acoustically Tagged Elasmobranchs." In *Shark Research: Emerging Technologies and Applications for the Field and Laboratory*. Edited by Jeffrey C. Carrier, Michael R. Heithaus, and Colin A. Simpfendorfer, 93–110. Boca Raton: Taylor & Francis. https://doi.org/10.1201/b21842

Lucchetti, Alessandro, Emilio Notti, Antonello Sala, and Massimo Virgili. 2018. "Multipurpose Use of Side-Scan Sonar Technology for Fisheries Science." *Canadian Journal of Fisheries and Aquatic Sciences* 75(10): 1652–1662. https://doi.org/10.1139/cjfas-2017-0359

Malakoff, David. 2002. "Suit Ties Whale Deaths to Research Cruise." *Science* 298(5594): 722–723. https://doi.org/10.1126/science.298.5594.722

Mallet, Robert. 1851. "Second Report on the Facts of Earthquake Phenomena." *Report of the British Association for the Advancement of Science*. 272–320. London: John Murray. https://www.biodiversitylibrary.org/page/13846657

Martins, A. P. B., M. R. Heupel, A. Oakley-Cogan, A. Chin, and C. A. Simpfendorfer. "Towed-Float GPS Telemetry: A Tool to Assess Movement Patterns and Habitat Use of Juvenile Stingrays." *Marine and Freshwater Research* 71(1): 89–98. https://doi.org/10.1071/MF19048

Mayer, Larry, Martin Jakobsson, Graham Allen, Boris Dorschel, Robin Falconer, Vicki Ferrini, Geoffroy Lamarche, Helen Snaith, and Pauline Weatherall. 2018. "The Nippon Foundation—GEBCO Seabed 2030 Project: The Quest to See the World's Oceans Completely Mapped by 2030." *Geosciences* 8(2): 63. https://doi.org/10.3390/geosciences8020063

McCauley, Robert. 2012. "Fish Choruses from the Kimberley, Seasonal and Lunar Links as Determined by Long Term Sea Noise Monitoring." In *Conference Proceedings of Acoustics 2012 Fremantle: Acoustics, Development and the Environment*. Edited by T. McMinn. Fremantle, Western Australia. November 21–23, 2012. http://hdl.handle.net/20.500.11937/45013

McMahon, Clive R., Fabien Roquet, Sophie Baudel, Mathieu Belbeoch, Sophie Bestley, Clint Blight, Lars Boehme, Fiona Carse, Daniel P. Costa, Michael A. Fedak, Christophe Guinet, Robert Harcourt, Emma Heslop, Mark A. Hindell, Xavier Hoenner, Kim Holland, Mellinda Hiolland, Fabrice R. A. Jaine, Tiphaine Jenniard du Dot, Ian Jonsen, Theresa R. Keates, Kit M. Kovacs, Sara Labrousse, Philip Lovel, Christian Lydersen, David March, Matthew Mazloff, Megan K. McKinzie, Mônica M.C. Muelbert, Kevin O'Brien, Lachlan Phillips, Esther Portela, Jonathan Pye, Stephen Rintoul, Katsufumi Sato, Ana M. M. Sequiera, Samantha E. Simmons, Vardis M. Tsontos, Victor Turpin, Esmee van Wijk, Danny Vo, Mia Wege, Frederick Gilbert Whoriskey, Kenady Wilson, and Bill Woodward. 2021. "Animal Borne Ocean Sensors – AniBOS – An Essential Component of the Global Ocean Observing System." *Frontiers in Marine Science* 8: 751840. https://doi.org/10.3389/fmars.2021.751840

McWilliam, Jamie N., Robert D. McCauley, Christine Erbe, and Miles J. G. Parsons. 2017. "Patterns of Biophonic Periodicity on Coral Reefs in the Great Barrier Reef." *Scientific Reports* 7: 17459. https://doi.org/10.1038/s41598-017-15838-z

Munk, Walter, and Carl Wunsch. 1979. "Ocean Acoustic Tomography: A Scheme for Large Scale Monitoring." *Deep Sea Research Part A. Oceanographic Research Papers* 26(2): 123–161. https://doi.org/10.1016/0198-0149(79)90073-6

Nandi, Papia, W. Steven Holbrook, Scott Pearse, Pedro Páramo, and Raymond W. Schmitt. 2004. "Seismic Reflection Imaging of Water Mass Boundaries in the Norwegian Sea." *Geophysical Research Letters* 31(23): L23311. https://doi.org/10.1029/2004GL021325

Nakamura, Y., T. Noguchi, T. Tsuji, S. Itoh, H. Niino, and T. Matsuoka. 2006. "Simultaneous Seismic Reflection and Physical Oceanographic Observations of Oceanic Fine Structure in the Kuroshio Extension Front." *Geophysical Research Letters* 33(23): L23605. https://doi.org/10.1029/2004GL021325

National Research Council. 1994. *Low-Frequency Sound and Marine Mammals: Current Knowledge and Research Needs*. Washington, DC: The National Academies Press. https://doi.org/10.17226/4557

National Research Council. 2000. *Marine Mammals and Low-Frequency Sound: Progress Since 1994*. Washington, DC: The National Academies Press. https://doi.org/10.17226/9756

National Research Council. 2003. *Ocean Noise and Marine Mammals*. Washington, DC: The National Academies Press. https://doi.org/10.17226/10564

Oleson, John Peter. 2000. "Ancient Sounding-Weights: A Contribution to the History of Mediterranean Navigation." *Journal of Roman Archaeology* 13: 293–310. https://doi.org/10.1017/S1047759400018948

Payne, Roger S., and Scott McVay. 1971. "Songs of Humpback Whales." *Science* 173(3997): 585–597. https://doi.org/10.1126/science.173.3997.585

Peschko, V., M. Mercker, and S. Garthe. 2020. "Telemetry Reveals Strong Effects of Offshore Wind Farms on Behaviour and Habitat Use of Common Guillemots (*Uria aalge*) During the Breeding Season." *Marine Biology* 167: 118. https://doi.org/10.1007/s00227-020-03735-5

Phillips, J. D., and D. F. Dean. 1991. "Multichannel Acoustic Reflection Profiling of Ocean Watermass Temperatures/Salinity Interfaces." In *Ocean Variability & Acoustic Propagation*. Edited by J. Potter and A. Warn-Varnas. 199–214. Dordrecht: Springer. https://doi.org/10.1007/978-94-011-3312-8_15

Pijanowski, Bryan C. , Almo Farina, Stuart H. Gage, Sarah L. Dumyahn, and Bernie L. Krause. "What Is Soundscape Ecology? An Introduction and Overview of an Emerging New Science." *Landscape Ecology* 26: 1213–1232. https://doi.org/10.1007/s10980-011-9600-8

Potter, John R. 1994. "ATOC: Sound Policy or Enviro-Vandalism? Aspects of a Modern Media-Fueled Policy Issue." *Journal of Environment & Development* 3(2): 47–62. https://doi.org/10.1177%2F107049659400300205

Ragland, John, Felix Schwock, Matthew Munson, and Shima Abadi. 2022. "An Overview of Ambient Sound Using Ocean Observatories Initiative Hydrophones." *The Journal of the Acoustical Society of America* 151: 2085–2100. https://doi.org/10.1121/10.0009836

Receveur, Aurore, Elodie Kestenare, Valerie Allain, Frédéric Ménard, Sophie Cravatte, Anne Lebourges-Dhaussy, Patrick Lehodey, Morgan Mangeas, Neville Smith, Marie-Hélène Radenac, and Christophe Menkes. 2020. "Micronekton Distribution in the Southwest Pacific (New Caledonia) Inferred from Shipboard-ADCP Backscatter Data." *Deep Sea Research Part I: Oceanographic Research Papers* 159: 103237. https://doi.org/10.1016/j.dsr.2020.103237

Rothenberg, David. 2014. "Nature's Greatest Hit: The Old and New Songs of the Humpback Whale." Wire. September 2014. https://www.thewire.co.uk/in-writing/essays/nature_s-greatest-hit_the-old-and-new-songs-of-the-humpback-whale

Ruddick, Barry, Haibin Song, Chingzhi Dong, and Luis Pinheiro. 2015. "Water Column Seismic Images as Maps of Temperature Gradient." *Oceanography* 22(1): 192–205. https://doi.org/10.5670/oceanog.2009.19

Savini, Alessandra. 2011. "Side-Scan Sonar as a Tool for Seafloor Imagery: Examples from the Mediterranean Continental Margin." In *Sonar Systems*. Edited by N.Z. Kolev. London: Intech Open. https://doi.org/ 10.5772/18375

Schafer, R. Murray. 1969. *The New Soundscape: A Handbook for the Modern Music Teacher*. Don Mills: BMI Canada Ltd. https://monoskop.org/images/0/03/Schafer_R_Murray_The_New_Soundscape_A_Handbook_for_the_Modern_Music_Teacher.pdf

Shang, Xiadong, Jianhu Zhao, and Hongmei Zhang. 2019. “Obtaining High-Resolution Seabed Topography and Surface Details by Co-Registration of Side-Scan Sonar and Multibeam Echo Sounder Images.” *Remote Sensing* 11(12): 1496. https://doi.org/10.3390/rs11121496

Sharpless, Andrew. 2012. “A Deaf Whale Is a Dead Whale.” USA Today. October 25, 2012. https://www.usatoday.com/story/opinion/2012/10/25/whales-deaf-air-gun-environment/1642207/

Slabbekoorn, Hans, Niels Bouton, Ilse van Opzeeland, Aukje Coers, Carel ten Cate, and Arthur N. Popper. 2010. “A Noisy Spring: The Impact of Globally Rising Underwater Sound Levels on Fish.” *Trends in Ecology & Evolution* 25(7): 419–427. https://doi.org/10.1016/j.tree.2010.04.005

Sousa-Lima, Renata S., Thomas F. Norris, Julie N. Oswald, and Deborah P. Fernandes. 2013. “A Review and Inventory of Fixed Autonomous Recorders for Passive Acoustic Monitoring of Marine Mammals.” *Aquatic Mammals* 39(1): 23–53. https://doi.org/10.1578/AM.39.1.2013.23

Southworth, Michael. 1969. “The Sonic Environment of Cities.” *Environment and Behavior* 1(1): 49–70. https://doi.org/10.1177/001391656900100104

Spence, Heather, Kyle M. Becker, and E. C. M. Parsons. 2022. “The Ocean Decade Research Program on the Maritime Acoustic Environment (OD-MAE).” *Marine Technology Society Journal* 56(3): 39–43. https://doi.org/10.4031/MTSJ.56.3.9

Staaterman, Erica, Claire B. Paris, Harry A. DeFerrari, David A. Mann, Aaron N. Rice, and Evan K. Ď Alessandro. 2014. “Celestial Patterns in Marine Soundscapes.” *Marine Ecology Progress Series* 508: 17–32. https://doi.org/10.3354/meps10911

Surlykke, Annemarie, and Paul E. Nachtigall. 2014. “Biosonar of Bats and Toothed Whales: An Overview.” In *Biosonar*. Edited by Annemarie Surlykke, Paul E. Nachtigall, Richard R. Fay, and Arthur N. Popper, 1–9. New York: Springer-Verlag. https://doi.org/10.1007/978-1-4614-9146-0_1

Tournadre, J. 2014. “Anthropogenic Pressure on the Open Ocean: The Growth of Ship Traffic Revealed by Altimeter Data Analysis.” *Geophysical Research Letters* 41: 7924–7932. https://doi.org/10.1002/2014GL061786

Treasure, Anne M., Fabien Roquet, Isabelle J. Ansorge, Marthán N. Bester, Lars Boehme, Horst Bornemann, Jean-Benoît, Damien Chevallier, Daniel P. Costa, Mike A. Fedak, Christophe Guinet, Mike O. Hammil, Robert G. Harcourt, Mark A. Hindell, Kit M. Kovacs, Mary-Anne Lea, Phil Lovell, Andrew D. Lowther, Christian Lydersen, Trevor McIntyre, Clive R. McMahon, Mônica M.C. Muelbert, Keith Nicholls, Baptiste Picard, Gilles Reverdin, Andrew W. Trites, Guy D. Williams, and P. J. Nico de Bruyn. 2017. “Marine Mammals Exploring the Oceans Pole to Pole: A Review of the MEOP Consortium.” *Oceanography* 30(2): 132–138. https://doi.org/10.5670/oceanog.2017.234

Tsuchiya, Toshio, Jun Naoi, Koji Futa, and Toshiaka Kikuchi. 2004. "Difference in Simulated Low-Frequency Sound Propagation in the Various Species of Baleen Whale." *Japanese Journal of Applied Physics* 43(5S): 3193. https://doi.org/10.1143/JJAP.43.3193

Tyack, Peter, George Frisk, Ian Boyd, Ed Urban, and Sophie Seeyave. 2015. *International Quiet Ocean Experiment Science Plan*. International Quiet Ocean Experiment. https://www.iqoe.org/products

van der Knapp, Inge, Jan Reubens, Len Thomas, Michael A. Ainslie, Hendrick V. Winter, Jeroen Hubert, Bruce Martin, and Hans Slabbekoorn. 2021. "Effects of a Seismic Survey on Movement of Free-Ranging Atlantic Cod." *Current Biology* 31(7): 1555–1562.e4 https://doi.org/10.1016/j.cub.2021.01.050

van der Knapp, Inge, Erin Ashe, Dave Hannay, Asila Ghoul Bergman, Kimberly A. Nelsen, Catherine F. Lo, and Rob Williams. 2022. "Behavioural Responses of Wild Pacific Salmon and Herring to Boat Noise." *Marine Pollution Bulletin* 174: 113257. https://doi.org/10.1016/j.marpolbul.2021.113257

Walsh, Emily A., John B. Kirkpatrick, Scott D. Rutherford, David C. Smith, Mitchell Sogin, and Steven D'Hondt. 2015. "Bacterial Diversity and Community Composition from Sea Surface to Subseafloor." *ISME Journal* 10: 979–989. https://doi.org/10.1038/ismej.2015.175

Weatherall, Pauline, K.M. Marks, Martin Jakobsson, Thierry Schmitt, Shin Tani, Jan Erik Arndt, Marzia Rovere, Dale Chayes, Vicki Ferrini, and Rochelle Wigley. 2015. "A New Digital Bathymetric Model of the World's Oceans." *Earth and Space Science* 2(8): 331–345. https://doi.org/10.1002/2015EA000107

Weatherby, B. B. 1940. "The History and Development of Seismic Prospecting." *Geophysics* 5(3): 215–230. https://doi.org/10.1190/1.1441805

Weilgart, Lindy. 2017. "Din of the Deep: Noise in the Ocean and Its Impacts on Cetaceans." In *Marine Mammal Welfare: Human Induced Change in the Marine Environment and its Impacts on Marine Mammal Welfare*. Edited by Andy Butterworth. Cham: Springer International Publishing. https://doi.org/10.1007/978-3-319-46994-2_7

Weiss, Kenneth R. 2002. "Research Involving Sonic Blasts Halted." Los Angeles Times. October 29, 2002. https://www.latimes.com/archives/la-xpm-2002-oct-29-me-whales29-story.html

Wiebe, Peter H., Timothy K. Stanton, Charles H. Greene, Mark C. Benfield, Heidi M. Sosik, Thomas C. Austin, Joseph D. Warren, and Terry Hammar. 2002. "BIOMAPER-II: An Integrated Instrument Platform for Coupled Biological and Physical Measurements in Coastal and Oceanic Regimes." *IEEE Journal of Oceanic Engineering* 27(3): 700–716. https://doi.org/10.1109/JOE.2002.1040951

Wiggins, Sean M., and John A. Hildebrand. 2020. "Fin Whale 40-Hz Calling Behavior Studied with an Acoustic Tracking Array." *Marine Mammal Science* 36(3): 964–971. https://doi.org/10.1111/mms.12680

Williams, Hannah J., Lucy A. Taylor, Simon Benhamou, Allert I. Bijleveld, Thomas A. Clay, Sophie de Grissac, Urška Demšar, Holly M. English, Novella Franconi, Agustina Gómez-Laich, Rachael C. Griffiths, William P. Kay, Juan Manuel Morales, Jonathan R. Potts, Katharine F. Rogerson, Christian Rutz, Anouk Spelt, Alice M. Trevail, Rory P. Wilson, and Luca Börger. 2020. "Optimizing the Use of Biologgers for Movement Ecology Research" *Journal of Animal Ecology* 89(1): 186–206. https://doi.org/10.1111/1365-2656.13094

Williams, R., A. J. Wright, E. Ashe, L. K. Blight, R. Bruintjes, R. Canessa, C. W. Clark, S. Cullis-Suzuki, D. T. Dakin, C. Erbe, P. S. Hammond, N. D. Merchant, P. D. O'Hara, J. Purser, A. N. Radford, S. D. Simpson, L. Thomas, and M. A. Wale. 2015. "Impacts of Anthropogenic Noise on Marine Life: Publication Patterns, New Discoveries, and Future Directions in Research and Management." *Ocean & Coastal Management* 115: 17–24. https://doi.org/10.1016/j.ocecoaman.2015.05.021

Wölfl, Anne-Cathrin, Helen Snaith, Sam Amirebrahimi, Colin W. Devey, Boris Dorschel, Vicki Ferrini, Veerle A. I. Huvenne, Martin Jakobsson, Jennifer Jencks, Gordon Johnston, Geoffrey Lamarche, Larry Mayer, David Millar, Terje Haga Pedersen, Kim Picard, Anja Reitz, Thierry Schmitt, Martin Visbeck, Pauline Weatherall, and Rochelle Wigley. 2019. "Seafloor Mapping – The Challenge of a Truly Global Ocean Bathymetry." *Frontiers in Marine Science* 6:283. https://doi.org/10.3389/fmars.2019.00283

Wu, Wenbo, Zhingwen Zhan, Shirui Peng, Sidao Ni, and Jörn Callies. 2020. "Seismic Ocean Thermometry ." *Science* 369(6510): 1510–1515. https://doi.org/10.1126/science.abb9519

Zhang, Boyu, Xiao Wang, Shudong Li, and Jinghua Yang. 2021. "Side Scan Sonar Shipwreck Image Recognition Algorithm based on Transfer Learning and Deep Learning." 3S *Web of Conferences* 290: 02020. https://doi.org/10.1051/e3sconf/202129002020

Zollinger, Sue Anne, and Henrik Brumm. 2011. "The Lombard Effect." *Current Biology* 21(16): PR14–PR15. https://doi.org/10.1016/j.cub.2011.06.003

FIGURE 5•79
If whales had headphones.

Credit: Shaw and Chamberlin

END VOYAGE I

Red lobster.

Artwork by Martha Rich.

Appendices

A. Common Abbreviations

4K 4,000 pixels
8K 8,000 pixels
16K 16,000 pixels
AI artificial intelligence
AU astronomical unit
BCE Before Common Era
CC Creative Commons
e.g. for example (*exempli gratia* in Latin)
et al. and others (*et alia* in Latin)
HD high definition
HMS His or Her Majesty's Ship
i.e. that is (*id est* in Latin)
ROSL roll on seafloor laughing
R/V or RV research vessel
SI International System of Units
SUV sport utility vehicle
TV television
UC University of California
UN United Nations
US United States
USS United States Ship

B. Ocean Science Abbreviations

A

ADS atmospheric diving suit
ALAMO Air-Launched Autonomous Micro-Observer
Argo Array for Real-Time Geostrophic Oceanography
ATOC Acoustic Thermometry of Ocean Climate
AUV autonomous underwater vehicle
AVHRR Advanced Very-High-Resolution Radiometers

C

CalCOFI California Cooperative Oceanic Fisheries Investigation
Conshelf Continental Shelf Station
CTD conductivity-temperature-depth
CZCS Coastal Zone Color Scanner instrument

D

DSV deep submergence vehicle
D-TAG digital acoustic recording tag

G

GDP global drifter program
GEBCO General Bathymetric Chart of the Oceans
GOES Geostationary Operational Environmental Satellite
GPS global positioning system

H

H-D hypothetico-deductive
HOV human-occupied vehicle
HROV hybrid remotely operated vehicle

I

IFREMER French Research Institute for Exploitation of the Sea

J

JAMSTEC Japan Agency for Marine-Earth Science and Technology

L

LED light-emitting diode

M

M-AUE mini-autonomous underwater explorer
MODIS Moderate Resolution Imaging Spectroradiometer
MOSAiC Multidisciplinary drifting Observatory for the Study of Arctic Climate
MUR-SSTs Multi-scale Ultra-high-Resolution sea surface temperatures

N

NASA National Aeronautics and Space Administration
NASEM National Academy of Sciences, Engineering, and Medicine
NDBC National Data Buoy Center
NeXOS Next generation, Cost-effective, Compact, Multifunctional Web Enabled Ocean Sensor Systems

NOAA National Oceanic and Atmospheric Administration
NRC National Research Council
NSF National Science Foundation

O

OFP Ocean Flux Program
OOI Ocean Observatories Initiative

P

PMEL Pacific Marine Environmental Laboratory
PSAT pop-up satellite archival tag

R

ROV remotely operated vehicle

S

SAR synthetic aperture radar
SAS synthetic aperture sonar
SCICEX SCientific ICe EXercise
SciComm science communication
Seasat Seafaring Satellite
SeaWiFS Sea-viewing Wide Field-of-view Sensor
SIO Scripps Institution of Oceanography
SOFAR Sound Fixing and Ranging
SOOP Ships of Opportunity Program
SOSUS Sound Surveillance System
SST sea surface temperature
Stratollite a portmanteau of stratosphere and satellite
SVP surface velocity program

T

TIROS Television Infrared Observation Satellite
TOPEX Ocean Topography Experiment

U

UNOLS University–National Oceanographic Laboratory System
USCG United States Coast Guard
USGS United States Geological Survey
USV uncrewed surface vessel
UUV uncrewed underwater vehicles

V

VIIRS Visible and Infrared Imager/Radiometer Suite

W

WHOI Woods Hole Oceanographic Institution

X

XBT expendable bathythermograph

C. Metric Prefixes

Prefix	Symbol	Name	Power of 10	Multiplier
yotta	Y	1 septillion	10^{24}	1,000,000,000,000,000,000,000,000
zetta	Z	1 sextillion	10^{21}	1,000,000,000,000,000,000,000
exa	E	1 quintillion	10^{18}	1,000,000,000,000,000,000
peta	P	1 quadrillion	10^{15}	1,000,000,000,000,000
tera	T	1 trillion	10^{12}	1,000,000,000,000
giga	G	1 billion	10^{9}	1,000,000,000
mega	M	1 million	10^{6}	1,000,000
kilo	k	1 thousand	10^{3}	1,000
hecto	h	1 hundred	10^{2}	1,00
deka	da	1 ten	10^{1}	10
deci	d	1 tenth	10^{-1}	0.1
centi	c	1 hundredth	10^{-2}	0.01
milli	m	1 thousandth	10^{-3}	0.001
micro	µ	1 millionth	10^{-6}	0.000,001
nano	n	1 billionth	10^{-9}	0.000,000,001
pico	p	1 trillionth	10^{-12}	0.000,000,000,001
femto	f	1 quadrillionth	10^{-15}	0.000,000,000,000,001
atto	a	1 quintillionth	10^{-18}	0. 000,000,000,000,000,001
zepto	z	1 sextillionth	10^{-21}	0. 000,000,000,000,000,000,001
yocto	y	1 septillionth	10^{-24}	0. 000,000,000,000,000,000,000,001

D. Useful Formulas

Length Conversions

to miles (mi)	=	[number of] km	×	0.62
	=	[number of] nmi	÷	1.151
	=	[number of] ft	÷	5,280
to kilometers (km)	=	[number of] mi	×	1.61
	=	[number of] nmi	×	1.852
	=	[number of] ft	÷	3,281
to nautical miles (nmi)	=	[number of] mi	÷	1.51
	=	[number of] km	÷	1.852
	=	[number of] ft	÷	6,076
to feet (ft)	=	[number of] mi	×	5,280
	=	[number of] km	×	3,281
	=	[number of] nmi	×	6,076

D. Useful Formulas (continued)

Area Conversions

to square miles (mi^2)	=	[number of] km^2	÷	2.59
	=	[number of] nmi^2	×	1.324
to square kilometers (km^2)	=	[number of] mi^2	×	2.59
	=	[number of] nmi^2	×	3.43
to square nautical miles (nmi^2)	=	[number of] mi^2	÷	1.324
	=	[number of] km^2	÷	3.43

Volume Conversions

to gallons (gal)	=	[number of] m^3	×	264.2
	=	[number of] L	÷	3.785
to cubic meters (m^3)	=	[number of] gal	÷	264.2
	=	[number of] L	÷	1,000
to liters (L)	=	[number of] gal	×	3.785
	=	[number of] m^3	×	1,000

Pressure Conversions

to atmospheres (atm)	=	[number of] mb	÷	1,013
	=	[number of] in. Hg	÷	29.92
	=	[number of] lb/in^2	÷	14.7
	=	[number of] Pa	÷	101,300
to millibars (mb)	=	[number of] atm	×	1,013
	=	[number of] in. Hg	×	33.864
	=	[number of] lb/in^2	×	68.948
	=	[number of] Pa	÷	100
to inches of Mercury (in. Hg)	=	[number of] atm	×	29.92
	=	[number of] mb	÷	33.864
	=	[number of] lb/in^2	×	2.036
	=	[number of] Pa	÷	3,386
to Pascals (Pa)	=	[number of] atm	×	101,300
	=	[number of] mb	×	100
	=	[number of] in. Hg	×	3,386
	=	[number of] lb/in^2	×	6,895

Speed Conversions

to miles per hour (mph)	=	[number of] knots	×	1.151
	=	[number of] km/hr	÷	1.609
	=	[number of] ft/s	÷	1.467
	=	[number of] m/s	×	2.237
to nautical miles per hour (knots)	=	[number of] mph	÷	1.151
	=	[number of] km/hr	÷	1.852
	=	[number of] ft/s	÷	1.688
	=	[number of] m/s	×	1.944
to kilometers per hour (km/hr)	=	[number of] mph	×	1.609
	=	[number of] knots	×	1.852
	=	[number of] ft/s	×	1.097
	=	[number of] m/s	×	3.6
to feet per second (f/s)	=	[number of] mph	×	1.467
	=	[number of] km/hr	÷	1.097
	=	[number of] knots	×	1.688
	=	[number of] m/s	×	3.281
to meters per second (m/s)	=	[number of] mph	÷	2.237
	=	[number of] knots	÷	1.944
	=	[number of] km/hr	÷	3.6
	=	[number of] ft/s	÷	3.281

Temperature

To convert degrees Fahrenheit (°F) to degrees Celsius (°C): (°F – 32) ÷ 1.8

To convert degrees Celsius (°C) to degrees Fahrenheit (°F): (°C × 1.8) + 32)

E. Estimating Water Pressure

Equation 3.1 (Chapter 3, p. 80) provides a convenient approximation for the calculation of water pressure in units of atmospheres from water depth in meters. Simply put, divide the depth (m) by 10 and add 1 atmosphere. Formally:

$$P\ (atm) = z\ (m)\ /\ 10\ (m/atm) + 1\ (atm)$$

(***Eq. 3.1***)

where P is water pressure, z is depth, and 1 atm represents the pressure at sea level.

For example, if the water depth is 80 meters, you would first divide by 10 (m/atm). The result, 8 atm, gives the water pressure for the water alone. But we must also take into account the force of the atmosphere, 1 atm. Adding 1 atm to 8 atm gives the correct answer of 9 atm for the water pressure at 80 meters. This is a reasonable estimate of water pressure at depth.

However, other factors, especially variations in the density of seawater, may affect water pressure. To more accurately calculate water pressure (formally known as hydrostatic pressure), you may wish to use the hydrostatic equation.

$$P\ (Pa) = \rho\ (kg\ m^{-3}) \times g\ (m\ s^{-2}) \times z\ (m)$$

(***Eq. A.1***)

where P is water pressure in Pascals (Pa, equivalent to Newtons per m^2), ρ (pronounced "row") is the density of the fluid in kilograms per cubic meter ($kg\ m^{-3}$), and g is the acceleration due to gravity, a constant at 9.8 m -2), and z is water depth in meters (m). Note that $1\ Pa = 1\ kg \times m^{-1} \times s^{-2}$ so the units work out (in case you were wondering).

The hydrostatic equation comes into play in water columns composed of layers with different densities. Physical oceanographers will use a calculus version of the equation (we're not going there), but for instructional purposes, just imaging three layers each with a slightly different density all stacked on top of each other. To accurately determine water pressure, you would need to calculate the pressure for each layer and add them up.

Now, unless you are pursuing a degree in oceanography, physics, or engineering, it's not likely you'll need to be familiar with the hydrostatic equation. But having a look beneath the hood, so to speak, will give you a greater appreciation for the measurement of pressure, on which so many things in the ocean depend. Should you ever descend in a submarine, you can rest assured the people who built it had a good understanding of this equation.

Deep-sea diving suit; istock/TomBaggett

Glossary

A

abductive reasoning coming to a conclusion based on experience, especially where data are limited.
acoustic niche a partition of the timing, intensity, or spectrum of sounds in a community of organisms to avoid overlap in sound-making.
acoustic tag an animal-attached device that transmits an acoustic signal that can be detected by a hydrophone.
acoustical oceanography the use and application of sound and its properties to study the ocean.
active acoustics the use of artificially produced sounds to study ocean properties and detect underwater objects.
active microwave sensor sensor that transmits and receives electromagnetic radiation to make measurements of Earth's surface.
active sonar a range-finding tool that produces a pulse of sound to locate an object through its echo.
activity model of the scientific method the interactive, back-and-forth, collaborative approach of teams of scientists for discovering knowledge about the natural world.
adaptive tracking system artificial intelligence–enhanced computer software that enables AUVs to navigate according to environmental conditions.
aerial investigation a scientific investigation that employs sensors aboard crewed or uncrewed airborne platforms to make observations or obtain measurements of natural processes.
air–sea interface the boundary between the ocean and the atmosphere.
amplitude the height of a wave crest above the undisturbed baseline.
animal oceanographer marine animal employed to carry sensors for oceanographic studies.
animal-tagging investigation a scientific investigation that uses sensors attached to animals to obtain observations or measurements of natural processes, generally in places inaccessible to humans.
anthrophony sounds produced by human vocalizations and activities.
anthroposphere the human system.
Argos satellite system a system of satellites that provide a data uplink for tags and other types of environmental monitoring devices.
artificial intelligence a broad category of computer science that aims to mimic how humans think, act, and adapt.
artificial seismology the use of human-generated seismic waves for studying Earth's interior.
atmosphere the gases surrounding Earth.
atmospheric diving suit a self-contained, pressurized, one-person diving suit used to explore waters deeper than scuba permits.
audible detectable by the human ear.
autonomous profiling float sensor-enabled robotic device that rises and sinks in the water column to carry out various ocean measurements and transmit the data to shore stations using satellites
autonomous underwater vehicle a self-propelled, self-navigating, sensor-equipped underwater vehicle for collecting oceanographic data or conducting search-and-rescue operations in the ocean.

B

bathymetric chart a map of a region of the ocean on which water depths are provided.
bathyscaphe a hollow steel sphere attached to a giant, gasoline-filled balloon for exploring the ocean depths.
bathysphere a hollow steel sphere with transparent baseball-sized portholes for making observations in the deep ocean.
benthic lander autonomous, sensor-enabled platform that sinks to the seafloor, carries out observations and experiments, and returns to the surface when signaled or at a predetermined time.

bias a less than objective, partial view of evidence.
big data a subdiscipline of computer science for analyzing data sets whose size or complexity exceeds the capabilities of traditional software and computers.
bio-optical oceanographer a human, robot, or animal that carries out optical measurements of the ocean's living systems.
bio-optical oceanography the scientific study of the optical properties of living systems in the ocean.
biologging the recording of data by devices attached to or embedded within an organism.
biological oceanography the scientific study of the biological properties and processes in the world ocean.
bionic AUV autonomous underwater vehicle that mimics the appearance and swimming behavior of a marine organism (also known as a robofish).
biophony sounds originating from biological sources (except humans).
biosphere the Earth's living systems.
biotelemetry the tracking of organisms using devices that continuously transmit signals to a receiver.
buoyancy the tendency of an immersed object to rise, sink, or remain in place.

C

chain of reasoning a logical extension of previous scientific work on a problem.
chemical oceanography the scientific study of the chemical properties and processes in the world ocean.
chlorophyll the primary light-absorbing pigment of all oxygen-producing photosynthetic organisms.
climate the long-term average of weather conditions at a given location or globally.
climate zone a region of Earth defined by a
coastal array ocean observatory deployed on a continental shelf (i.e., near a coastline).
computer modeling investigation a scientific investigation that uses computers to visualize or quantify natural processes.
conceptual model a mental picture, often in visual form, of how we think something works.
conduction the transfer of heat through molecule-**coral bleaching** the expulsion of photosynthetic symbionts by a coral polyp in response to prolonged elevated water temperatures.
crest the high point of a sine wave.
cryosphere the frozen parts of Earth.
cycle the completion of one compression–expansion motion of air molecules as a sound wave moves through a medium.

D

deductive reasoning the approach of eliminating all possible hypotheses until only one cannot be disproven.
deep scattering layer a layer (or layers) where high concentrations of vertically migrating organisms alter the propagation of sound.
diel vertical migration the daily upward-and-downward movement of marine organisms.
digital acoustic recording tag animal-attached tag that records sound and other environmental data.
deep circulation the three-dimensional flows of **diving bell** dome-shaped, air-filled chamber that permits divers to work at depth.
drift bottle method the use of buoyant containers for tracking ocean currents.
drifter a platform that floats on the ocean's surface to track currents.

E

Earth system the various interdependent, interconnected systems that collectively regulate Earth.
echolocation the use of sound to locate and identify objects
ecosystem service an economic, ecological, restorative, or cultural benefit provided by an ecosystem to humans.
emergent property a property of a system not apparent when studying its parts alone.
expendable bathythermograph a disposable device for recording seawater temperature as it changes with depth in the water column.

F

falsification the process of disproving a hypothesis.
fantail the back end (i.e., stern) of a ship.
fathom a unit of depth; about the length of a person's outstretched arms.
fathometer an acoustic device for determining water depth.
fish chorus sustained and elevated sound-making by fishes for an hour or more.
field investigation *See* oceangoing investigation.
float a platform that drifts beneath the ocean's surface to track currents.

G

geological oceanography the scientific study of the geological properties and processes in the world ocean.
geophone microphone that listens for the sound vibrations caused by seismic waves on land.
geophony sounds emanating from geologic and physical processes.
geophysics the study of solid Earth through observations of its physical properties.
geophere the solid Earth.
glider winged, sensor-enabled, buoyancy-driven platform for carrying out various measurements over large expanses of the ocean.
global array ocean observatory deployed in a deep water location.
global climate change the change in Earth's climate over geologic time.
global warming the increase in Earth's average temperature since 1850.
GPS tag animal-attached tag that records the geographic location (i.e., latitude, longitude, elevation) of the animal.
ground truthing the process of independently verifying the accuracy and precision of measurements made from satellites.

H

hadal lander benthic lander capable of reaching the deepest parts of the world ocean.
heliosphere the Sun and solar system.
hertz a unit of sound defined as one cycle per second.
holey sock drogue a large, subsurface cloth cylinder attached to a surface float that improves the accuracy of current measurements.
hybrid remotely operated vehicle a cross between an ROV and an AUV that permits operation with or without a tether.
hydrographic station a shipboard sampling location in the ocean, usually arranged as points on a grid.
hydrosphere Earth's waters.
hyperspectral imager sensor capable of detecting hundreds of narrow bands of electromagnetic radiation.
hypothesis a carefully worded, conditional, and testable explanation of how nature works.
hypothetico-deductive approach the general process of generating and disproving hypotheses based on observations or experiments.

I

iceberg a chunk of glacier floating in the ocean.
inductive reasoning the process of extracting general principles and possible hypotheses from a limited set of observations or data.
infrasonic too low to be detected by the human ear.
interdisciplinary science a science that integrates the tools, approaches, knowledge, and understanding of multiple disciplines to solve scientific problems.

K

kilohertz a unit of sound defined as 1,000 cycles per second.
kite aerial photography a type of above-surface photography that uses kites to elevate the camera.

L

laboratory investigation a scientific investigation carried out in a laboratory.
Lagrangian platform a platform that moves with the currents.

M

manipulator arm human-controlled, mechanical extension on an ROV for collecting samples, drilling holes, or deploying instruments in the ocean.
marine big data the enormous volumes of oceanographic data now available and growing daily.
mathematical models simple or complex mathematical expressions or sets of expressions to describe or simulate how something works.
mesophotic coral ecosystems coral reefs that grow in dimly lit waters.
mini-autonomous underwater explorer small, autonomous, sensor- and GPS-enabled platform that works in a swarm beneath the surface of the ocean to track currents or carry out other kinds of observations and measurements.
mowing the lawn technique the technique whereby a ship travels back and forth across the ocean's surface to map the seafloor in the same manner that a person mows a lawn.
multibeam bathymetry seafloor depth chart produced by multibeam echosounders.
multibeam echosounder device that employs multiple acoustic sources to obtain depth information over a wide patch of the seafloor.
multispectral imager sensor capable of detecting multiple wide bands of electromagnetic radiation.

N

noise any unwanted or undesired sound in the environment.
nowcast forecast of ocean or atmospheric conditions in near real time.
numerical models computer models that solve sets of mathematical equations to make forecasts or projections of natural processes at some time in the future.

O

ocean acoustic tomography the use of low-frequency sound to observe ocean temperatures.
ocean bottom hydrophone underwater microphone deployed on the seafloor.
ocean bottom seismometer instrument for detecting and recording seismic waves on the seafloor.
ocean color the colors of visible light emanating from the surface of the ocean.
ocean eddy a self-sustaining, rotating loop of water tens to hundreds of miles wide.
ocean mooring seafloor-anchored cable on which various ocean sensors can be attached and held vertically by a float.
ocean observatory an integrated array of ocean-observing sensors and platforms on and above the seafloor designed to address questions of scientific and societal importance.
ocean-observing systems a broad category of sensor-outfitted platforms that fly above, glide on top of, or dive below the surface of the ocean.
oceangoing investigation a scientific investigation carried out at sea.
oceanographer any person, robot, or animal engaged in the scientific study of the ocean.
oceanography the scientific study of the world ocean.
offshore the region beyond the surf zone.
operational oceanography the acquisition and dissemination of oceanographic and meteorological information useful to navigating or carrying out activities in the ocean.

P

pale blue dot the name given to the blue speck of Earth as seen from millions of miles away in outer space.
passive acoustic recorder device that combines modern electronics with digital recording devices to obtain sound data for long periods of time.
passive acoustics the study of underwater sounds using hydrophones.
passive microwave sensor sensor that measures the intensity of microwave radiation emanating from Earth's land, ice, and sea surface.
passive sensor sensor that measures the electromagnetic properties of the ocean.
peer review the process whereby scientists receive feedback on their work by other scientists and experts.

physical model a scaled-down, three-dimensional, semirealistic representation of a physical object or part of nature.
physical oceanography the scientific study of the physical properties and processes in the world ocean.
physical state the state of matter: solid, liquid, or **phytoplankton** photosynthetic microbes that drift with the currents in the sunlit waters of the ocean.
pop-up satellite archival tag a sensor-enabled animal tag that releases from the animal after a period of time and floats to the surface to upload its data
pressure the force exerted by a fluid (a gas or a liquid) on an object immersed within it.
principle of unity the idea that there is but one ocean on Earth, the world ocean.
progressive wave a wave moving forward along a density boundary.

Q

qualitative data described by its properties or relationships with other variables.
quantitative numerical data that can be analyzed mathematically.

R

radiometer sensor that detects electromagnetic radiation.
rarefaction expansion of air molecules as a sound wave moves through a medium
refraction the bending of waves in response to the properties of the medium through which they pass
regional cabled array seafloor-based ocean observatory spanning a region of geologic interest.
remotely operated vehicle human-controlled, camera-equipped, propelled vehicle attached to a shipboard tether for gathering images, making measurements, or collecting samples from the water column or seafloor.
robotic investigation a scientific investigation that employs robotic devices to make observations and measurements of the natural world or under artificially created conditions.

S

saildrone a wind-propelled, solar-powered surfboard to which instruments may be attached for gathering oceanographic and meteorological information.
satellite altimeter an active microwave sensor used to determine the bumps and depressions of the sea surface.
satellite oceanography the use of Earth-orbiting satellites to make observations and collect data on ocean features and processes.
satellite product any of a number of computed variables derived from measurements using satellites.
science communication the widespread and transparent dissemination of scientific knowledge and understanding, especially in a manner accessible to the public.
science literacy the knowledge and understanding of science required to participate meaningfully in personal and societal discussions and decisions that use science.
scientific discussion the presentation in oral or written form of the conclusions of a particular scientific study.
scientific diving the use of scuba diving to carry out scientific investigations underwater.
scientific inquiry a general term to describe the diverse approaches scientists use to investigate the natural world.
scientific method the systematic and self-correcting set of practices that scientists use to discover knowledge about the natural world.
scientific question a well-formed question that lends itself to investigation using the scientific method.
scuba an acronym for self-contained underwater breathing apparatus.

sea surface height the elevation of the sea surface relative to Earth's geoid—the bumps and depressions of the sea surface.
seismic airgun a ship-towed device that emits explosive bangs of compressed air to create vibrations for seismic studies in the ocean.
seismic oceanography the use of seismic reflection methods for detecting the thermal structure (i.e., thermoclines) of the water column.
seismic profiling the use of artificially produced seismic waves to map the three-dimensional structure of Earth's crust.
seismic prospecting the use of seismic methods for petroleum and mineral exploration.
seismic reflection profile a two- or three-dimensional image of the subsurface obtained using seismic reflection methods.
seismic reflection profiling detection of the reflected echoes of artificially produced seismic waves that reveal the structure of rock layers beneath Earth's surface.
seismic refraction profile a two- or three-dimensional image of the subsurface obtained using seismic refraction methods.
seismic refraction profiling detection of the refracted echoes of artificially produced seismic waves that reveal the structure of rock layers beneath Earth's surface.
seismic wave motion of the ground caused by the movement of energy through Earth's crust.
seismology the study of earthquakes.
seismometer a device for detecting and recording the passage of seismic waves.
shark cage a metal cage that permits safe viewing of sharks and other pelagic ocean species.
ship of opportunity private, commercial, or nongovernmental ship enlisted to take oceanographic and meteorological measurements on its voyage.
side-scan sonar an acoustic device that transmits high-frequency sound pulses toward the seafloor at low angles to produce high-resolution images of seafloor features or sunken objects.
sine wave the up-and-down graphical representation of a trigonometric sine function
sinusoidal relating to or shaped like a sine wave.
SOFAR channel *See* Sound Fixing and Ranging channel.
sonar an acronym for sound navigation and ranging; typically refers to acoustic systems used to find underwater objects.
song vocalization by a marine mammal that repeats in distinct patterns.
sonic *See* audible.
sonograph near-photographic-quality image obtained from side-scan sonar.
SOSUS See Sound Surveillance system.
Sound Fixing and Ranging channel the depth range encompassing the sound minimum layer which permits sound waves to travel great distances in the ocean.
sound minimum layer the subsurface region of the ocean where the speed of sound reaches a minimum.
sound scattering layer *See* deep scattering layer.
Sound Surveillance system a system of hydrophones established by the US Navy to detect enemy submarines.
sound vibrations you can hear or detect with instruments.
sounding line a rope to which a weight is attached for determining water depth.
sounding the depth determined using a sounding line, fathometer, or other device.
soundscape the totality of sounds present in a particular region or habitat.
soundscape ecology the study of the behavior and social interactions of sound makers, including humans
spaceborne investigation a scientific investigation that employs sensors aboard Earth-orbiting platforms to make observations and obtain measurements of Earth's surface.
spatial resolution relating to units of space (distance, area, volume).
steerable stratospheric balloon a high-altitude balloon capable of staying aloft for long periods of time.

streamer the name given to the towed array of hydrophones behind a ship conducting seafloor seismic studies.
submarine a human-occupied vessel capable of propelling itself underwater and remaining submerged for long periods of time.
submersible a small submarine capable of reaching the deepest depths of the ocean.
Swallow float a subsurface platform with acoustic transmitters for tracking ocean currents.
synthetic aperture radar an active microwave sensor used to generate two- and three-dimensional images of Earth's surface, especially sea ice.
synthetic aperture sonar a type of sonar that averages multiple signals to obtain a highly detailed image of an object on the seafloor.
system an assemblage of parts and processes working together.

T

telepresence a livestream from a ship-based ROV transmitted via satellite that permits remote users to watch live video and interact with scientists and technicians on board.
temporal resolution relating to units of time, from fractions of seconds to millennia.
tether a metal-strengthened fiber-optic cable that transmits two-way signals between an operator and an ROV.
thermocline a layer of the water column in which temperature declines rapidly.
thumper truck vehicle with a device that pounds the ground to create vibrations for seismic studies on land.
time-series investigation a scientific investigation that observes and measures natural processes over time.
tipping point a change in the state of a system that is unstoppable or irreversible.
tomol wood plank canoes constructed by the Chumash people
tow-yo pattern a W-shaped pattern of sampling using a towfish alternately raised and lowered on its path.
towfish a winged, sensor-enabled platform towed behind a vessel on a cable.
transect a line of stations on a grid.
trough the low point of a sine wave.
twilight zone a zone characterized by dim and diffuse light.

U

ultrasonic too high to be detected by the human ear.
uncrewed aerial platform uncrewed, human-controlled, or autonomous platform that carries out missions in Earth's atmosphere.
uncrewed surface vessel a self-propelled, self-navigating, sensor-equipped vehicle carrying out various missions on the surface of the ocean.
uncrewed underwater vehicle a self-propelled, self-navigating, sensor-equipped vehicle carrying out various missions beneath the ocean.
undersampling problem the lack of sufficient data to fully understand the ocean over all scales of space and time.
undersea research station semipermanent undersea structure designed for human habitation and scientific research on the seafloor.
underwater swarm robotics an emerging field of robotics in which multiple autonomous platforms interact with each other to solve problems or carry out missions.

W

wave frequency the number of cycles in a fixed interval of time (usually a second).
wave glider a sensor-enabled platform that uses solar and wave power to glide through the ocean.
wave period the time between successive crests of a progressive wave.
wave refraction the bending of a wave front in response to changes in seafloor bathymetry.
wave speed the distance a wave travels over a given period of time.
wavelength the horizontal distance between successive crests in a wave.
weather the instantaneous state of the atmosphere.
world ocean all of the waters contained within the ocean basins.

Index

Potato cod (Epinephelus tukula); istock/Volkantg

About the Authors

W. Sean Chamberlin, PhD is an American oceanographer, professor, and author. He holds bachelor's degrees in English and Oceanography from the University of Washington, and a PhD in Biology from the University of Southern California. Since 1996 he has taught oceanography, meteorology, and climate science at Fullerton College in Fullerton, California (a few short miles from Disneyland). Chamberlin has traveled four of the five major oceans (sorry, Indian Ocean) and carried out studies aboard Jacques Cousteau's *Calypso*. He co-authored *Exploring the World Ocean* (Chamberlin and Dickey 2008) and has published numerous papers in ocean science and online education. When not teaching or writing, he cares for three cats and spends much-treasured time with his beloved partner of 22+ years. Professor Sean's website

Thank you for reading our book.

Nicole Shaw is an American cartoonist, writer, business analyst, and mother to four kids. She holds a BS in Bioengineering from UC San Diego. In 2017, she started @nickisdoodles, an Instagram feed featuring fun science facts and puppy comics. She currently works as a senior business analyst for a medical IT company in Durham, North Carolina. When not doodling or making computers behave, she helps her husband run their specialty coffee business, Olilu Coffee. Nicki's website

Martha Rich is an American artist, illustrator, and professor. She is a graduate of (and has taught classes at) the Art Center College of Design in Pasadena, California, and holds an MFA from the University of Pennsylvania. She has also taught at Tyler School of Art and Architecture and the Fashion Institute of Technology. In 2019, Chronicle Books published her book, *The Furious Notebook, Release Your Rage, Use Your Anger for Good and Chill the Heck Out.* Her work has appeared in music videos, books, magazines, murals and in various products sold worldwide. In 2023, a collection of her work appeared at the Hey There Projects gallery in Joshua Tree, California. Martha's website